Model-Based Control

Paul M.J. Van den Hof • Carsten Scherer
Peter S.C. Heuberger
Editors

Model-Based Control

Bridging Rigorous Theory and Advanced Technology

 Springer

Editors
Paul M.J. Van den Hof
Delft Center for Systems and Control
Delft University of Technology
Mekelweg 2
2628 CD Delft
The Netherlands
p.m.j.vandenhof@tudelft.nl

Carsten Scherer
Delft Center for Systems and Control
Delft University of Technology
Mekelweg 2
2628 CD Delft
The Netherlands
c.w.scherer@tudelft.nl

Peter S.C. Heuberger
Delft Center for Systems and Control
Delft University of Technology
Mekelweg 2
2628 CD Delft
The Netherlands
p.s.c.heuberger@tudelft.nl

ISBN 978-1-4899-8372-5   ISBN 978-1-4419-0895-7 (eBook)
DOI 10.1007/978-1-4419-0895-7
Springer Dordrecht Heidelberg London New York

Printed on acid-free paper

Springer is part of Springer Science+Business Media (www.springer.com)

Foreword

This book is written at the occasion of the sixty-fifth birthday of Professor Okko Bosgra on June 13th, 2009, and commemorates the role that Okko Bosgra has played in developing and promoting the field of systems and control theory and engineering during a lifetime effort of more than forty years at Delft University of Technology.

At the occasion of this birthday, the Delft Center for Systems and Control (DCSC) at Delft University of Technology has taken the initiative to organize a one-day symposium "Four decades of systems and control - in honor of professor Okko Bosgra" that took place on September 18, 2009 in the Aula of Delft University of Technology. During this symposium several of Okko's colleagues and former students have contributed by giving a seminar.

This book collects research contributions of many coworkers and former students of Okko, who have been inspired and supported by him in looking for the challenging research questions beyond the current state of the art.

Acknowledgements

Many colleagues, friends and former students contributed to this book and the symposium. We thank them all, especially Ellen van den Berg-Moor and her team for her invaluable work in the organization of the symposium and the preparation of this book[1].

---

[1] The cover illustration of this book was reprinted from the cover of "Selected Topics in Modelling, Identification and Control", (O.H. Bosgra, P.M.J. Van den Hof and C.W. Scherer (eds)), published from 1990-2001, with permission from IOS Press and from the designer Ruud Schrama.

After his graduation in 1968 at Delft University of Technology, Professor Okko
H. Bosgra started his career as an assistant and associate professor in the measurement and control group of the Mechanical Engineering Department. After a part-
time professorship of four years at the Department of Physics at Wageningen University (1981-1985), he returned to Delft to take up a full professorship in Systems
and Control in the same Mechanical Engineering Department.

Okko was strongly driven towards establishing a research group which covers
the whole breadth of the field, ranging from modeling and identification to optimal
and robust control, at the forefront of international developments. Although being

located in an application oriented environment, he always had a high esteem of fundamental mathematically oriented research in control. During the period of extension of his research group (1986-1999), he very successfully created an environment in which his vision of high-quality research in control could flourish, namely to allow for a close bond between theoretical developments at the forefront of science and technological questions resulting from the confrontation with industrial practice. Testimony of the achievements are the variety of the topics covered in the large number of PhD- and MSc-theses supervised by Okko, in addition to having earned the highest scores after research evaluations by an international committee of peers in the years 1994 and 2000. During this period and under his leadership the group issued the twelve volume journal "Selected Topics in Modeling, Identification and Control" in which some of the obtained results were collected in order to increase their dissemination and to boost visibility of the group's output.

In the second half of the eighties Okko has played an important role in the emerging national cooperation between the mathematical system theory groups and the control engineering groups. This strong collaborative initiative of which Okko was a firm supporter from the early beginning, is the backbone of the national institute DISC (Dutch Institute of Systems and Control), which has obtained international recognition through its established national PhD programme. First as member of the board and later as scientific director (2000-2005), Okko has played an important role in shaping and directing this national cooperation.

During this period Okko also created strong links to various Dutch industries which provided an early and steady stream of financial support. A particularly strong relationship was established with the mechatronic and motion control research department at Philips in Eindhoven for which he has been serving as a scientific advisor since 1987. Motivated by the needs of high-tech industries like ASML, the world-leading manufacturer of wafer scanners, Okko promoted a combination of the control and mechatronics activities in Delft to a new department which he headed in 2000-2004. During this period he was a driving force behind the preparation to merge three control groups in Delft spread over three faculties into the Delft Center for Systems and Control. Since 2005 Okko holds a part-time professorship at Delft, in addition to part-time professorships in the Control Systems Technology Group of the Mechanical Engineering Department and the Electrical Engineering Department of Eindhoven University of Technology.

Education of students is a matter dear to Okko's heart. For more than twenty years he has been teaching advanced control courses both at the MSc and PhD level. His excellent role of supervisor and mentor of students was recognized by Delft University of Technology in 1997 when he was awarded the University Award for Excellence in Teaching ("Leermeesterprijs"). In parallel to the creation of the Delft Center for Systems and Control, he was among the initiators for installing the first MSc program in Systems and Control in the Netherlands in 2003.

Much of Okko's effort was devoted to servicing the university and the Dutch community in all aspects related to scientific developments and in various capacities. Among others he held positions in the Dutch Foundation for Post-Academic Technical Education, the Royal Institute of Engineers in the Netherlands, the Microned consortium, and the European Union Control Association in parallel to which he acted as an Editor at Large of the European Journal of Control (1994-1999). During his career more than 400 MSc students and 60 PhD students have graduated under his supervision, many of whom, clearly marked as his scholars, found their way in industry and academia.

Okko, this book is dedicated to you, by all editors and authors who have enjoyed working with you and being inspired by you, with great appreciation for your role and support.

Delft,   *Paul M.J. Van den Hof*
April, 2009   *Carsten W. Scherer*
                                                                               Peter S.C. Heuberger

Preface

The first part of this volume is a collection of four papers on the fundamental aspects of control. The article by Willems is devoted to an exposition of the behavioral approach to linear discrete-time dynamical systems, with a particular emphasis on developing various representations of sets of trajectories as images and kernels of difference operators. Scherer's paper discusses convex optimization techniques for robust controller synthesis if the underlying generalized plant description has a particular structure and if the uncertainties are captured by integral quadratic constraints. Modeling of physical systems in different domains by conservations laws is the topic of the third paper by Van der Schaft and Maschke in which the authors show how to move from two- to higher-dimensional system networks on the basis of the theory of k-complexes from algebraic topology. The last contribution by Dreesen and de Moor centers around the role of multi-variable polynomial optimization in systems theory and presents algorithms how to solve such problems by realization theory and eigenvalue computations.

In the second part we collected those articles that bridge theoretical developments with technologically advanced applications. In the contribution of Li, de Oliveira and Skelton the authors reveal how to include the selection of sensors and actuators in an convex optimization-based optimal feedback design algorithm, with an illustration of controller synthesis for a flexible beam. Balas, Packard and Seiler suggest linear matrix inequality techniques for modeling frequency-domain uncertainties as they are required in robust controller design, with an application to data sets obtained for the NASA Generic Transport Model aircraft. Viewing feedback controller tuning as a problem of robust estimation is considered by Kinney and de Callafon. They show how their approach can be employed for real-time controller adjustment based on closed-loop measurements and how it applies to an experimental setup for active noise suppression. In the fourth article Van den Hof, Van Doren and Douma are concerned with parameter identifiability if modeling large-scale physical systems on the basis of measurement data, such as required for optimal oil recovery from large reservoirs.

The third part of the book collects articles devoted to applications in motion control and process engineering. Steinbuch, van de Wijdeven, Oomen, van Berkel and Leenknegt suggest an iterative learning algorithm in order to handle cracks on discs if trying to extract information by optical drive systems. Their experimental results for a Blue Ray device confirm the resulting benefits for recovering data from damaged discs. The article by Tousain and van der Meulen looks into the optimal on-line determination of feed-forward control structures which permits the adaptation to real-time variations in the plant dynamics. Experimental results are described for a wafer-stage and for a digital light projection system in a commercial product. The last two articles address dynamic process control. Kahrs, Brendel, Michalik and Marquardt present their incremental approach to model identification which involves the decomposition of a large system into smaller subsystems for reducing complexity. The modeling approach is demonstrated by experimental results for a semi-batch reactor with a homogeneous liquid reactor content. The book is concluded by a contribution from Jansen, Yortsos, Van Doren and Heidary-Fyrozjaee on the limited controllability of sharp front waves of immiscible two-phase flows in homogenous media. For reservoir engineering they conclude that the presence of heterogeneities in the permeability field is an essential ingredient in order to influence subsurface fluid flows through well-rate manipulation.

Contents

Part III Applications in Motion Control Systems and Industrial Process Control

Part IV Appendix

List of Contributors

Gary J. Balas
Aerospace Engineering and Mechanics, University of Minnesota, Minneapolis,
MN 55455, U.S.A., e-mail: balas@umn.edu

Marc Brendel
Aachener Verfahrenstechnik - Process Systems Engineering, RWTH Aachen
University, Germany. Current affiliation: Evonik Degussa GmbH, Hanau, Germany.
e-mail: marc.brendel@evonik.com

Raymond A. de Callafon
Department of Mechanical and Aerospace Engineering, University of California,
San Diego, 9500 Gilman Drive, La Jolla, CA 92093-0411, U.S.A., e-mail:
callafon@ucsd.edu

Bart De Moor
Katholieke Universiteit Leuven, Department of Electrical Engineering –
ESAT/SCD, Kasteelpark Arenberg 10, 3001 Leuven, Belgium. e-mail:
bart.demoor@esat.kuleuven.be

Maurício C. de Oliveira
Department of Mechanical and Aerospace Engineering, University of California
San Diego, La Jolla, CA 92093-0411, U.S.A., e-mail: mauricio@ucsd.edu

Sippe G. Douma
Shell International Exploration and Production, P.O. Box 60, 2280 AB Rijswijk,
The Netherlands, e-mail: sippe.douma@shell.com

Philippe Dreesen
Katholieke Universiteit Leuven, Department of Electrical Engineering –
ESAT/SCD, Kasteelpark Arenberg 10, 3001 Leuven, Belgium. e-mail:
philippe.dreesen@esat.kuleuven.be

Mohsen Heidary-Fyrozjaee
University of Southern California, Viterbi School of Engineering, 3650 McClintock
Ave, Los Angeles, CA 90089 USA, e-mail: heidarif@usc.edu

Jan Dirk Jansen
Delft University of Technology, Department of Geotechnology, PO box 5048, 2600
GA, Delft, The Netherlands, and Shell International E&P, Kessler Park 1, 2288 GS
Rijswijk, The Netherlands, e-mail: j.d.jansen@tudelft.nl

Olaf Kahrs
Aachener Verfahrenstechnik - Process Systems Engineering, RWTH Aachen
University, Germany. Current affiliation: BASF SE, Ludwigshafen, Germany.
e-mail: olaf.kahrs@basf.com

Charles E. Kinney
Department of Mechanical and Aerospace Engineering, University of California,
San Diego, 9500 Gilman Drive, La Jolla, CA 92093-0411, U.S.A., e-mail:
cekinney@ucsd.edu

George Leenknegt
Advanced Research Center, Philips Lite-On Digital Solutions Netherlands, Glaslaan
2, 5616 LW Eindhoven, The Netherlands, e-mail: george.leenknegt@pldsnet.com

Faming Li
Xerox Research Center, Webster, NY, U.S.A., e-mail: Faming.Li@XEROX.com

Wolfgang Marquardt
Aachener Verfahrenstechnik - Process Systems Engineering, RWTH Aachen
University, Germany, e-mail: wolfgang.marquardt@avt.rwth-aachen.de

Bernhard M. Maschke
Lab. d'Automatique et de Genie des Procédés, Université Claude Bernard Lyon-1,
F-69622 Villeurbanne, Cedex, France, e-mail: maschke@lagep.univ-lyon1.fr

Claas Michalik
Aachener Verfahrenstechnik - Process Systems Engineering, RWTH Aachen
University, Germany,

Tom Oomen
Eindhoven University of Technology, Department of Mechanical Engineering,
Control Systems Technology group, P.O. Box 513, 5600 MB Eindhoven, The
Netherlands, e-mail: t.a.e.oomen@tue.nl

Andrew K. Packard
Mechanical Engineering, University of California, Berkeley, CA 94720, U.S.A,
e-mail: apackard@berkeley.edu

Carsten W. Scherer
Delft Center for Systems and Control, Delft University of Technology, Mekelweg
2, 2628 CD Delft, The Netherlands, e-mail: c.w.scherer@tudelft.nl

Peter J. Seiler
Aerospace Engineering and Mechanics, University of Minnesota, Minneapolis,
MN 55455, U.S.A., e-mail: seiler@aem.umn.edu

Robert E. Skelton
Department of Mechanical and Aerospace Engineering, University of California
San Diego, La Jolla, CA 92093-0411, U.S.A., e-mail: bobskelton@ucsd.edu

Maarten Steinbuch
Eindhoven University of Technology, Department of Mechanical Engineering,
Control Systems Technology group, P.O. Box 513, 5600 MB Eindhoven, The
Netherlands, e-mail: m.steinbuch@tue.nl

Rob Tousain
Philips Lighting, Mathildelaan 1, 5611 BD, Eindhoven, The Netherlands, e-mail:
rob.tousain@philips.com

Koos van Berkel
Eindhoven University of Technology, Department of Mechanical Engineering,
Control Systems Technology group, P.O. Box 513, 5600 MB Eindhoven, The
Netherlands, e-mail: k.v.berkel@gmail.com

Paul M.J. Van den Hof
Delft Center for Systems and Control,Delft University of Technology, Mekelweg 2,
2628 CD Delft, The Netherlands, e-mail: p.m.j.vandenhof@tudelft.nl

Stan van der Meulen
Eindhoven University of Technology, Department of Mechanical Engineering,
Control Systems Technology Group, PO Box 513, 5600 MB, Eindhoven, The
Netherlands, e-mail: S.H.v.d.Meulen@tue.nl

Arjan van der Schaft
Institute of Mathematics and Computing Science, University of Groningen, PO Box
407, 9700 AK, Groningen, The Netherlands, e-mail: A.J.van.der.Schaft@rug.nl

Jeroen van de Wijdeven
TMC Mechatronics, P.O. Box 700, 5600 AS Eindhoven, The Netherlands, e-mail:
jeroen.van.de.wijdeven@tmc.nl

Jorn F.M. Van Doren
Delft Center for Systems and Control, Delft University of Technology, Mekelweg
2, 2628 CD Delft, The Netherlands, and Shell International E&P, Kessler Park 1,
2288 GS Rijswijk, The Netherlands, e-mail: jorn.vandoren@shell.com

Jan C. Willems
ESAT-SISTA, K.U. Leuven, B-3001 Leuven, Belgium, e-mail:
Jan.Willems@esat.kuleuven.be

Yannis C. Yortsos
University of Southern California, Viterbi School of Engineering, 3650 McClintock
Ave, Los Angeles, CA 90089 USA, e-mail: yortsos@usc.edu

Part I
Fundamentals

Linear Systems in Discrete Time

Jan C. Willems

Abstract Representations of linear time-invariant discrete-time systems are discussed. A system is defined as a behavior, that is, as a family of trajectories mapping the time axis into the signal space. The following characterizations are equivalent: (i) the system is linear, time-invariant, and complete, (ii) the behavior is linear, shift-invariant, and closed, (iii) the behavior is kernel of a linear difference operator with a polynomial symbol, (iv) the system allows a linear input/output representation in terms of polynomial matrices, (v) the system allows a linear constant coefficient input/state/output representation, and (vi) the behavior is kernel of a linear difference operator with a rational symbol. If the system is controllable, then the system also allows (vii) an image representation with a polynomial symbol, and an image representation with a rational symbol.

1 Introduction

It is a pleasure to contribute an article to this Festschrift dedicated to Professor Okko Bosgra on the occasion of his 'emeritaat'.

The aim of this presentation is to discuss representations of discrete-time linear time-invariant systems described by difference equations. We discuss systems from the behavioral point of view. Details of this approach may be found in [1, 2, 3, 4, 5].

We view a model as a subset $\mathscr{B}$ of a universum $\mathscr{U}$ of a priori possibilities. This subset $\mathscr{B} \subseteq \mathscr{U}$ is called the *behavior* of the model. Thus, before the phenomenon was captured in a model, all outcomes from $\mathscr{U}$ were in principle possible. But after we accept $\mathscr{B}$ as the model, we declare that only outcomes from $\mathscr{B}$ are possible.

In the case of *dynamical* systems, the phenomenon which is modeled produces functions that map the set of time instances relevant to the model to the *signal space*.

Jan C. Willems
ESAT-SISTA, K.U. Leuven, B-3001 Leuven, Belgium e-mail: Jan.Willems@esat.kuleuven.be

P.M.J. Van den Hof et al. (eds.), *Model-Based Control: Bridging Rigorous Theory and Advanced Technology*, DOI: 10.1007/978-1-4419-0895-7_1,
© Springer Science + Business Media, LLC 2009

This is the space in which these functions take on their values. In this article we assume that the set of relevant time instances is $\mathbb{N} := \{1, 2, 3, \ldots\}$ (the theory is analogous for $\mathbb{Z}, \mathbb{R}$, and $\mathbb{R}_+$). We assume also that the signal space is a finite-dimensional real vector space, typically $\mathbb{R}^{\mathtt{w}}$.

Following our idea of a model, the behavior of the dynamical systems which we consider is therefore a collection $\mathscr{B}$ of functions mapping the time set $\mathbb{N}$ into the signal space $\mathbb{R}^{\mathtt{w}}$. A dynamical model can therefore be identified with its *behavior* $\mathscr{B} \subseteq (\mathbb{R}^{\mathtt{w}})^{\mathbb{N}}$. The behavior is hence a family of maps from $\mathbb{N}$ to $\mathbb{R}^{\mathtt{w}}$. Of course, also for dynamical systems the behavior $\mathscr{B}$ is usually specified as the set of solutions of equations, for the case at hand typically difference equations. As dynamical models, difference equations thus merely serve as a representation of their solution set. Note that this immediately leads to a notion of equivalence and to canonical forms for difference equations. These are particularly relevant in the context of dynamical systems, because of the multitude of, usually over-parameterized, representations of the behavior of a dynamical system.

2 Linear dynamical systems

The most widely studied model class in systems theory, control, and signal processing consists of dynamical systems that are (i) linear, (ii) time-invariant, and (iii) that satisfy a third property, related to the finite dimensionality of the underlying state space, or to the rationality of a transfer function. It is, however, clearer and advantageous to approach this situation in a more intrinsic way, by imposing this third property directly on the behavior, and not on a representation of it. The purpose of this presentation is to discuss various representations of this model class.

A behavior $\mathscr{B} \subseteq (\mathbb{R}^{\mathtt{w}})^{\mathbb{N}}$ is said to be *linear* if $w \in \mathscr{B}, w' \in \mathscr{B}$, and $\alpha \in \mathbb{R}$ imply $w + w' \in \mathscr{B}$ and $\alpha w \in \mathscr{B}$, and *time-invariant* if $\sigma \mathscr{B} \subseteq \mathscr{B}$. The *shift* σ is defined by $(\sigma f)(\mathtt{t}) := f(\mathtt{t} + 1)$. The third property that enters into the specification of the model class is completeness. $\mathscr{B}$ is called *complete* if it has the following property:

$$[\![w : \mathbb{N} \to \mathbb{R}^{\mathtt{w}} \text{ belongs to } \mathscr{B}]\!] \Leftrightarrow [\![w|_{[1,\mathtt{t}]} \in \mathscr{B}|_{[1,\mathtt{t}]} \text{ for all } \mathtt{t} \in \mathbb{N}]\!].$$

In words, $\mathscr{B}$ is complete if we can decide that $w : \mathbb{N} \to \mathbb{R}^{\mathtt{w}}$ is 'legal' (i.e. belongs to $\mathscr{B}$) by verifying that each of its 'prefixes' $(w(1), w(2), \ldots, w(\mathtt{t}))$ is 'legal' (i.e. belongs to $\mathscr{B}|_{[1,\mathtt{t}]}$). So, roughly speaking, $\mathscr{B}$ is complete iff the laws of $\mathscr{B}$ do not involve what happens at $+\infty$. Requirements as $w \in \ell_2(\mathbb{N}, \mathbb{R}^{\mathtt{w}})$, w has compact support, or $\lim_{\mathtt{t} \to \infty} w(\mathtt{t})$ exists, risk at obstructing completeness. However, often crucial information about a complete $\mathscr{B}$ can be obtained by considering its intersection with $\ell_2(\mathbb{N}, \mathbb{R}^{\mathtt{w}})$, or its compact support elements, etc.

Recall the following standard notation. $\mathbb{R}[\xi]$ denotes the polynomials with real coefficients in the indeterminate ξ, $\mathbb{R}(\xi)$ the real rational functions, and $\mathbb{R}^{n_1 \times n_2}[\xi]$ the polynomial matrices with real $n_1 \times n_2$ matrices as coefficients. When the number of rows is irrelevant and the number of columns is n, the notation $\mathbb{R}^{\bullet \times n}[\xi]$ is used.

So, in effect, $\mathbb{R}^{\bullet \times n}[\xi] = \cup_{k \in \mathbb{N}} \mathbb{R}^{k \times n}[\xi]$. A similar notation is used for polynomial vectors, or when the number of rows and/or columns is irrelevant. The degree of $P \in \mathbb{R}^{\bullet \times \bullet}[\xi]$ equals the largest degree of its entries, and is denoted by **degree** (P).

Given a time-series $w : \mathbb{N} \to \mathbb{R}^w$ and a polynomial matrix $R \in \mathbb{R}^{v \times w}[\xi]$, say $R(\xi) = R_0 + R_1 \xi + \cdots + R_L \xi^L$, we can form the new v-dimensional time-series

$$R(\sigma)w = R_0 w + R_1 \sigma w + \cdots + R_L \sigma^L w.$$

Hence $R(\sigma) : (\mathbb{R}^w)^{\mathbb{N}} \to (\mathbb{R}^v)^{\mathbb{N}}$, with $R(\sigma)w : t \in \mathbb{N} \mapsto R_0 w(t) + R_1 w(t+1) + \cdots + R_L w(t+L) \in \mathbb{R}^v$.

The combination of linearity, time-invariance, and completeness can be expressed in many equivalent ways. In particular, the following are equivalent:

1) $\mathscr{B} \subseteq (\mathbb{R}^w)^{\mathbb{N}}$ is linear, time-invariant, and complete;
2) $\mathscr{B}$ is a linear, shift invariant $(:\Leftrightarrow \sigma \mathscr{B} \subseteq \mathscr{B})$, closed subset of $(\mathbb{R}^w)^{\mathbb{N}}$, with 'closed' understood in the topology of pointwise convergence;
3) $\exists R \in \mathbb{R}^{\bullet \times w}[\xi]$ such that $\mathscr{B}$ consists of the solutions $w : \mathbb{N} \to \mathbb{R}^w$ of

$$R(\sigma)w = 0. \tag{1}$$

The set of behaviors $\mathscr{B} \subseteq (\mathbb{R}^w)^{\mathbb{N}}$ that satisfy the equivalent conditions 1. to 3. is denoted by $\mathscr{L}^w$, or, when the number of variables is unspecified, by $\mathscr{L}^{\bullet}$. Thus, in effect, $\mathscr{L}^{\bullet} - \cup_{w \in \mathbb{N}} \mathscr{L}^w$. Since $\mathscr{B} = \textbf{kernel}(R(\sigma))$ in (1), we call (1) a *kernel representation* of the behavior $\mathscr{B}$.

3 Polynomial annihilators

We now introduce a characterization that is mathematically more abstract. It identifies a behavior $\mathscr{B} \in \mathscr{L}^{\bullet}$ with an $\mathbb{R}[\xi]$-module.

Consider $\mathscr{B} \in \mathscr{L}^w$. The polynomial vector $n \in \mathbb{R}^{1 \times w}[\xi]$ is called an *annihilator* (or a *consequence*) of $\mathscr{B}$ if $n(\sigma)\mathscr{B} = 0$, i.e. if $n(\sigma)w = 0$ for all $w \in \mathscr{B}$. Denote by $\mathscr{N}_{\mathscr{B}}$ the set of annihilators of $\mathscr{B}$. Observe that $\mathscr{N}_{\mathscr{B}}$ is an $\mathbb{R}[\xi]$-module. Indeed, $n \in \mathscr{N}_{\mathscr{B}}, n' \in \mathscr{N}_{\mathscr{B}}$, and $\alpha \in \mathbb{R}[\xi]$ imply $n + n' \in \mathscr{N}_{\mathscr{B}}$ and $\alpha n \in \mathscr{N}_{\mathscr{B}}$. Hence the map $\mathscr{B} \mapsto \mathscr{N}_{\mathscr{B}}$ associates with each $\mathscr{B} \in \mathscr{L}^w$ a submodule of $\mathbb{R}^{1 \times w}[\xi]$. It turns out that this map is actually a bijection, i.e. to each submodule of $\mathbb{R}^{1 \times w}[\xi]$, there corresponds exactly one element of $\mathscr{L}^w$. It is easy to see what the inverse map is. Let $\mathscr{K}$ be a submodule of $\mathbb{R}^{1 \times w}[\xi]$. Submodules of $\mathbb{R}^{1 \times w}[\xi]$ have nice properties. In particular, they are *finitely generated*, meaning that there exist elements ('*generators*') $g_1, g_2, \ldots, g_g \in \mathscr{K}$ such that $\mathscr{K}$ consists precisely of the linear combinations $\alpha_1 g_1 + \alpha_2 g_2 + \cdots + \alpha_g g_g$ where the α_k's range over $\mathbb{R}[\xi]$. Now consider the system (1) with $R = \text{col}(g_1, g_2, \ldots, g_g)$ and prove that

$$\mathscr{N}_{\textbf{kernel}(\text{col}(g_1, g_2, \ldots, g_g)(\sigma))} = \mathscr{K}$$

($\supseteq$ is obvious, $\subseteq$ requires a little bit of analysis). In terms of (1), we obtain the characterization

$$[\,\mathbf{kernel}(R(\sigma)) = \mathscr{B}\,] \quad \Leftrightarrow \quad [\,\mathscr{N}_{\mathscr{B}} = \langle R\rangle\,]$$

where $\langle R\rangle$ denotes the $\mathbb{R}[\xi]$-module generated by the rows of R.

The observation that there is a bijective correspondence between $\mathscr{L}^{\mathtt{w}}$ and the $\mathbb{R}[\xi]$-submodules of $\mathbb{R}^{1\times\mathtt{w}}[\xi]$ is not altogether trivial. For instance, the surjectivity of the map

$$\mathscr{B} = \mathbf{kernel}(R(\sigma)) \in \mathscr{L}^{\mathtt{w}} \quad \mapsto \quad \mathscr{N}_{\mathscr{B}} = \langle R\rangle$$

onto the $\mathbb{R}[\xi]$-submodules of $\mathbb{R}^{1\times\mathtt{w}}[\xi]$ depends on the solution concept used in (1). If we would have considered only solutions with compact support, or that are square integrable, this bijective correspondence is lost. Equations, in particular difference or differential equations, all by themselves, without a clear solution concept, i.e. without a definition of the corresponding behavior, are an inadequate specification of a mathematical model. Studying linear time-invariant difference (and certainly differential) equations is not just algebra, through the solution concept, it also requires analysis.

The characterization of $\mathscr{B}$ in terms of its module of annihilators shows precisely what we are looking for in order to identify a system in the model class $\mathscr{L}^{\bullet}$: (a set of generators of) the submodule $\mathscr{N}_{\mathscr{B}}$.

4 Input/output representations

Behaviors in $\mathscr{L}^{\bullet}$ admit many other representations. The following two are exceedingly familiar to system theorists. In fact,

4) $[\mathscr{B} \in \mathscr{L}^{\mathtt{w}}] \Leftrightarrow [\exists$ integers $\mathtt{m},\mathtt{p} \in \mathbb{Z}_{+}$, with $\mathtt{m}+\mathtt{p} = \mathtt{w}$, polynomial matrices $P \in \mathbb{R}^{\mathtt{p}\times\mathtt{p}}[\xi], Q \in \mathbb{R}^{\mathtt{p}\times\mathtt{m}}[\xi]$, with $\det(P) \neq 0$, and a permutation matrix $\Pi \in \mathbb{R}^{\mathtt{w}\times\mathtt{w}}$ such that $\mathscr{B}$ consists of all $w : \mathbb{N} \to \mathbb{R}^{\mathtt{w}}$ for which there exist $u : \mathbb{N} \to \mathbb{R}^{\mathtt{m}}$ and $y : \mathbb{N} \to \mathbb{R}^{\mathtt{p}}$ such that

$$P(\sigma)y = Q(\sigma)u \qquad (2)$$

and $w = \Pi \begin{bmatrix} u \\ y \end{bmatrix}$]. The matrix of rational functions $G = P^{-1}Q \in (\mathbb{R}(\xi))^{\mathtt{p}\times\mathtt{m}}$ is called the *transfer function* of (2). Actually, for a given $\mathscr{B} \in \mathscr{L}^{\mathtt{w}}$, it is always possible to choose Π such that G is proper. If we would allow a basis change in $\mathbb{R}^{\mathtt{w}}$, i.e. allow any non-singular matrix for Π (instead of only a permutation matrix), then we could always take G to be strictly proper.

5) $[\mathscr{B} \in \mathscr{L}^{\mathtt{w}}] \Leftrightarrow [\exists$ integers $\mathtt{m},\mathtt{p},\mathtt{n} \in \mathbb{Z}_{+}$ with $\mathtt{m}+\mathtt{p} = \mathtt{w}$, matrices $A \in \mathbb{R}^{\mathtt{n}\times\mathtt{n}}, B \in \mathbb{R}^{\mathtt{n}\times\mathtt{m}}, C \in \mathbb{R}^{\mathtt{p}\times\mathtt{n}}, D \in \mathbb{R}^{\mathtt{p}\times\mathtt{m}}$, and a permutation matrix $\Pi \in \mathbb{R}^{\mathtt{w}\times\mathtt{w}}$ such that $\mathscr{B}$ consists of all $w : \mathbb{N} \to \mathbb{R}^{\mathtt{w}}$ for which there exist $u : \mathbb{N} \to \mathbb{R}^{\mathtt{m}}, x : \mathbb{N} \to \mathbb{R}^{\mathtt{n}}$, and $y : \mathbb{N} \to \mathbb{R}^{\mathtt{p}}$ such that

$$\sigma x = Ax + Bu, \quad y = Cx + Du, \quad w = \Pi \begin{bmatrix} u \\ y \end{bmatrix}]. \tag{3}$$

If we allow also a basis change in $\mathbb{R}^w$, i.e. allow any non-singular matrix for Π, then we can also take $D = 0$.

(2) is called an *input/output* (i/o) and (3) an *input/state/output* (i/s/o) representation of the corresponding behavior $\mathscr{B} \in \mathscr{L}^w$.

Why, if any element $\mathscr{B} \in \mathscr{L}^\bullet$ indeed admits a representation (2) or (3), should one not use one of these familiar representations ab initio? There are many good reasons for not doing so. To begin with, and most importantly, first principles models aim at describing a behavior, but are seldom in the form (2) or (3). Consequently, one must have a theory that supersedes (2) or (3) in order to have a clear idea what transformations are allowed in bringing a first principles model into the form (2) or (3). Secondly, as a rule, physical systems are simply not endowed with a signal flow direction. Adding a signal flow direction is often a figment of one's imagination, and when something is not real, it will turn out to be cumbersome sooner or later. A third reason, very much related to the second, is that the input/output framework is totally inappropriate for dealing with all but the most special system interconnections. We are surrounded by interconnected systems, but only very sparingly can these be viewed as input-to-output connections. The second and third reason are valid, in an amplified way, for continuous-time systems. Fourthly, the structure implied by (2) or (3) often needlessly complicates matters, mathematically and conceptually.

A good theory of systems takes the behavior as the basic notion and the reference point for concepts and definitions, and switches back and forth between a wide variety of convenient representations. (2) or (3) have useful properties, but for many purposes other representations may be more convenient. For example, a kernel representation (1) is very relevant in system identification. It suggests that we should look for (approximate) annihilators. On the other hand, when it comes to constructing trajectories, (3) is very convenient. It shows how trajectories are parameterized and generated : by the initial state $x(1) \in \mathbb{R}^n$ and the input $u : \mathbb{N} \to \mathbb{R}^m$.

5 Representations with rational symbols

Our next representation involves rational functions and is a bit more 'tricky'. Let $G \in (\mathbb{R}(\xi))^{\bullet \times w}$ and consider the system of 'difference equations'

$$G(\sigma)w = 0. \tag{4}$$

What is meant by the behavior of (4)? Since G is a matrix of rational functions, it is not evident how to define solutions. This may be done in terms of co-prime factorizations, as follows. G can be factored $G = P^{-1}Q$ with $P \in \mathbb{R}^{\bullet \times \bullet}[\xi]$ square, $\det(P) \neq 0, Q \in \mathbb{R}^{\bullet \times w}[\xi]$ and (P,Q) left co-prime (meaning that $F = [P \ Q]$ is left prime, i.e.

$$[(U, F' \in \mathbb{R}^{\bullet \times \bullet}[\xi]) \wedge (F = UF')] \Rightarrow [U \text{ is square and unimodular }],$$

equivalently $\exists\, H \in \mathbb{R}^{\bullet \times \bullet}[\xi]$ such that $FH = I$). We *define* the behavior of (4) as that of

$$Q(\sigma)w = 0, \quad \text{i.e. as} \quad \mathbf{kernel}(Q(\sigma))$$

Hence (4) defines a behavior $\in \mathscr{L}^{\mathtt{w}}$. It is easy to see that this definition is independent of which co-prime factorization is taken. There are other reasonable ways of approaching the problem of defining the behavior of (4), but they all turn out to be equivalent to the definition given. Rational representations are studied in [6]. Note that, in a trivial way, since (1) is a special case of (4), every element of $\mathscr{L}^{\mathtt{w}}$ admits a representation (4).

6) $[\mathscr{B} \in \mathscr{L}^{\mathtt{w}}] \Leftrightarrow [\text{there exists } G \in \mathbb{R}(\xi)^{\bullet \times \mathtt{w}} \text{ such that it admits a kernel representation (4)}]$.

6 Integer invariants

Certain integer 'invariants' (meaning maps from $\mathscr{L}^{\bullet}$ to $\mathbb{Z}_+$) associated with systems in $\mathscr{L}^{\bullet}$ are important. One is the *lag*, denoted by $\mathsf{L}(\mathscr{B})$, defined as the smallest $\mathsf{L} \in \mathbb{Z}_+$ such that $[w|_{[\mathtt{t},\mathtt{t}+\mathsf{L}]} \in \mathscr{B}|_{[1,\mathtt{t}+1]}$ for all $\mathtt{t} \in \mathbb{N}] \Rightarrow [w \in \mathscr{B}]$. Equivalently, the smallest degree over the polynomial matrices R such that $\mathscr{B} = \mathbf{kernel}(R(\sigma))$. A second integer invariant that is important is the *input cardinality*, denoted by $\mathsf{m}(\mathscr{B})$, defined as $\mathtt{m}$, the number of input variables in any (2) representation of $\mathscr{B}$. It turns out that $\mathtt{m}$ is an invariant (while the input/output partition, i.e. the permutation matrix Π in (2), is not). The number of output variables, $\mathtt{p}$, yields the *output cardinality* $\mathsf{p}(\mathscr{B})$. A third important integer invariant is the *state cardinality*, $\mathsf{n}(\mathscr{B})$, defined as the smallest number $\mathtt{n}$ of state variables over all i/s/o representations (3) of $\mathscr{B}$. The three integer invariants $\mathsf{m}(\mathscr{B})$, $\mathsf{n}(\mathscr{B})$, and $\mathsf{L}(\mathscr{B})$ can be nicely captured in one single formula, involving the growth as a function of $\mathtt{t}$ of the dimension of the subspace $\mathscr{B}|_{[1,\mathtt{t}]}$. Indeed, there holds

$$\dim(\mathscr{B}|_{[1,\mathtt{t}]}) \leq \mathsf{m}(\mathscr{B})\,\mathtt{t} + \mathsf{n}(\mathscr{B}) \text{ with equality iff } \mathtt{t} \geq \mathsf{L}(\mathscr{B}).$$

7 Latent variables

State models (3) are an example of the more general, but very useful, class of latent variable models. Such models involve, in addition to the *manifest* variables (denoted by w in (5)), the variables which the model aims at, also auxiliary, *latent* variables (denoted by ℓ in (5)). For the case at hand this leads to behaviors $\mathscr{B}_{\text{full}} \in \mathscr{L}^{\mathtt{w}+1}$ described by

$$R(\sigma)w = M(\sigma)\ell, \tag{5}$$

with $R \in \mathbb{R}^{\bullet \times \mathtt{w}}[\xi]$ and $M \in \mathbb{R}^{\bullet \times 1}[\xi]$.

Although the notion of observability applies more generally, we use it here for latent variable models only. We call $\mathscr{B}_{\mathrm{full}} \in \mathscr{L}^{\mathtt{w}+1}$ *observable* if

$$[\![(w, \ell_1) \in \mathscr{B}_{\mathrm{full}} \text{ and } (w, \ell_2) \in \mathscr{B}_{\mathrm{full}}]\!] \Rightarrow [\![\ell_1 = \ell_2]\!].$$

(5) defines an observable latent variable system iff $M(\lambda)$ has full row rank for all $\lambda \in \mathbb{C}$. For state systems (with x the latent variable), this corresponds to the usual observability of the pair (A, C).

An important result, the *elimination theorem*, states that $\mathscr{L}^{\bullet}$ is closed under projection. Hence $\mathscr{B}_{\mathrm{full}} \in \mathscr{L}^{\mathtt{w}+1}$ implies that the *manifest* behavior

$$\mathscr{B} = \mathbf{projection}(\mathscr{B}) = \{w : \mathbb{N} \to \mathbb{R}^{\mathtt{w}} \mid \exists \ell : \mathbb{N} \to \mathbb{R}^1 \text{ such that (5) holds}\}$$

belongs to $\mathscr{L}^{\mathtt{w}}$, and therefore admits a kernel representation (1) of its own. So, in a trivial sense, (5) is yet another representation of $\mathscr{L}^{\mathtt{w}}$.

Latent variable representations (also unobservable ones) are very useful in all kinds of applications. This, notwithstanding the elimination theorem. They are the end result of modeling interconnected systems by *tearing, zooming, and linking* [5], with the interconnection variables viewed as latent variables. Many physical models (for example, in mechanics) express basic laws using latent variables.

8 Controllability

In many areas of system theory, controllability enters as a regularizing assumption. In the behavioral theory, an appealing notion of controllability has been put forward. It expresses what is needed intuitively, it applies to any dynamical system, regardless of its representation, it has the classical state transfer definition as a special case, and it is readily generalized, for instance to distributed systems. It is somewhat strange that this definition has not been generally adopted. Adapted to the case at hand, it reads as follows. The time-invariant behavior $\mathscr{B} \subseteq (\mathbb{R}^{\bullet})^{\mathbb{N}}$ is said to be *controllable* if for any $w_1 \in \mathscr{B}$, $w_2 \in \mathscr{B}$, and $\mathtt{t}_1 \in \mathbb{N}$, there exists a $\mathtt{t}_2 \in \mathbb{N}$ and a $w \in \mathscr{B}$ such that $w(\mathtt{t}) = w_1(\mathtt{t})$ for $1 \leq \mathtt{t} \leq \mathtt{t}_1$, and $w(\mathtt{t}) = w_2(\mathtt{t} - \mathtt{t}_1 - \mathtt{t}_2)$ for $\mathtt{t} > \mathtt{t}_1 + \mathtt{t}_2$. For $\mathscr{B} \in \mathscr{L}^{\bullet}$, one can take without loss of generality $w_1 = 0$ in the above definition. Denote the controllable elements of $\mathscr{L}^{\bullet}$ by $\mathscr{L}^{\bullet}_{\mathrm{cont}}$ and of $\mathscr{L}^{\mathtt{w}}$ by $\mathscr{L}^{\mathtt{w}}_{\mathrm{cont}}$.

The kernel representation (1) defines a controllable system iff $R(\lambda)$ has the same rank for each $\lambda \in \mathbb{C}$. There is a very nice representation result that characterizes controllability: it is equivalent to the existence of an image representation. More precisely, $\mathscr{B} \in \mathscr{L}^{\bullet}_{\mathrm{cont}}$ iff there exists $M \in \mathbb{R}^{\bullet \times \bullet}[\xi]$ such that $\mathscr{B}$ equals the manifest behavior of the latent variable system

$$w = M(\sigma)\ell. \tag{6}$$

In other words,

7) $[\![\mathscr{B} \in \mathscr{L}_{\text{cont}}^{\bullet}]\!] \Leftrightarrow [\![\mathscr{B} = \text{im}\,(M(\sigma))]\!]$.

So, images, contrary to kernels, are always controllable. This image representation of a controllable system can always be taken to be observable.

For $\mathscr{B} \in \mathscr{L}^{\bullet}$, we define its *controllable part*, denoted by $\mathscr{B}_{\text{controllable}}$, as

$$\mathscr{B}_{\text{controllable}} := \{w \in \mathscr{B} \mid \forall t' \in \mathbb{N}, \exists t'' \in \mathbb{Z}_+, \text{ and } w' \in \mathscr{B} \text{ such that}$$
$$w'(t) = 0 \text{ for } 1 \leq t \leq t' \text{ and } w'(t) = w(t - t' - t'') \text{ for } t > t' + t''\}.$$

Equivalently, $\mathscr{B}_{\text{controllable}}$ is the largest controllable subsystem contained in $\mathscr{B}$. It turns out that two systems of the form (2) (with the same input/output partition) have the same transfer function iff they have the same controllable part.

9 Rational annihilators

Consider $\mathscr{B} \in \mathscr{L}^{\mathtt{w}}$. The vector of rational functions $n \in \mathbb{R}^{1 \times \mathtt{w}}(\xi)$ is called a *rational annihilator* of $\mathscr{B}$ if $n(\sigma)\,\mathscr{B} = 0$ (note that, since we gave a meaning to (4), this is well defined). Denote by $\mathscr{N}_{\mathscr{B}}^{\text{rational}}$ the set of rational annihilators of $\mathscr{B}$. Observe that $\mathscr{N}_{\mathscr{B}}^{\text{rational}}$ is a $\mathbb{R}(\xi)$-subspace of $\mathbb{R}^{1 \times \mathtt{w}}(\xi)$. The map $\mathscr{B} \mapsto \mathscr{N}_{\mathscr{B}}^{\text{rational}}$ is not a bijection from $\mathscr{L}^{\mathtt{w}}$ to the $\mathbb{R}(\xi)$-subspaces of $\mathbb{R}^{1 \times \mathtt{w}}(\xi)$. Indeed,

$$[\![\,\mathscr{N}_{\mathscr{B}'}^{\text{rational}} = \mathscr{N}_{\mathscr{B}''}^{\text{rational}}\,]\!] \Leftrightarrow [\![\,\mathscr{B}_{\text{controllable}}' = \mathscr{B}_{\text{controllable}}''\,]\!].$$

However, there exists a bijective correspondence between $\mathscr{L}_{\text{cont}}^{\mathtt{w}}$ and the $\mathbb{R}(\xi)$-subspaces of $\mathbb{R}^{1 \times \mathtt{w}}(\xi)$. Summarizing, $\mathbb{R}[\xi]$-submodules of $\mathbb{R}^{1 \times \mathtt{w}}[\xi]$ stand in bijective correspondence with $\mathscr{L}^{\mathtt{w}}$, with each submodule corresponding to the set of polynomial annihilators, while $\mathbb{R}(\xi)$-subspaces of $\mathbb{R}^{1 \times \mathtt{w}}(\xi)$ stand in bijective correspondence with $\mathscr{L}_{\text{cont}}^{\mathtt{w}}$, with each subspace corresponding to the set of rational annihilators.

Controllability enters in a subtle way whenever a system is identified with its transfer function. Indeed, it is easy to prove that the system described by

$$w_2 = G(\sigma)w_1, \quad w = \begin{bmatrix} w_1 \\ w_2 \end{bmatrix}, \tag{7}$$

a special case of (4), is automatically controllable. This again shows the limitation of identifying a system with its transfer function. Two input/output systems (2) with the same transfer function are the same iff they are both controllable. In the end, transfer function thinking can deal with non-controllable systems only in contorted ways.

10 Stabilizability

A property related to controllability is stabilizability. The behavior $\mathscr{B} \subseteq (\mathbb{R}^\bullet)^{\mathbb{N}}$ is said to be *stabilizable* if for any $w \in \mathscr{B}$ and $t \in \mathbb{N}$, there exists a $w' \in \mathscr{B}$ such that $w'(t') = w(t')$ for $1 \le t' \le t$, and $w'(t) \to 0$ for $t \to \infty$. (1) defines a stabilizable system iff $R(\lambda)$ has the same rank for each $\lambda \in \mathbb{C}$ with Real$(\lambda) \ge 0$. An important system theoretic result (leading up to the parametrization of stabilizing controllers) states that $\mathscr{B} \in \mathscr{L}^{\mathtt{w}}$ is stabilizable iff it allows a representation (4) with $G \in (\mathbb{R}(\xi))^{\bullet \times \mathtt{w}}$ left prime over the ring $\mathbb{RH}_\infty$ ($:= \{f \in \mathbb{R}(\xi) \mid f$ is proper and has no poles in the closed right half of the complex plane $\}$). $\mathscr{B} \in \mathscr{L}^{\mathtt{w}}$ is controllable iff it allows a representation $w = G(\sigma)\ell$ with $G \in (\mathbb{R}(\xi))^{\mathtt{w} \times \bullet}$ right prime over the ring $\mathbb{RH}_\infty$.

11 Autonomous systems

Autonomous systems are on the other extreme of controllable ones. $\mathscr{B} \subseteq (\mathbb{R}^\bullet)^{\mathbb{N}}$ is said to be *autonomous* if for every $w \in \mathscr{B}$, there exists a $t \subset \mathbb{N}$ such that $w|_{[1,t]}$ uniquely specifies $w|_{[t+1,\infty)}$, i.e. such that $w' \in \mathscr{B}$ and $w|_{[1,t]} = w'|_{[1,t]}$ imply $w' = w$. It can be shown that $\mathscr{B} \in \mathscr{L}^\bullet$ is autonomous iff it is finite dimensional. Autonomous systems and, more generally, uncontrollable systems are of utmost importance in systems theory, in spite of much system theory folklore claiming the contrary. Controllability as a system property is much more restrictive than is generally appreciated.

Acknowledgments
The SISTA-SMC research program is supported by the Research Council KUL: GOA AM-BioRICS, CoE EF/05/006 Optimization in Engineering (OPTEC), IOF-SCORES4CHEM, several PhD/postdoc and fellow grants; by the Flemish Government: FWO: PhD/postdoc grants, projects G.0452.04 (new quantum algorithms), G.0499.04 (Statistics), G.0211.05 (Nonlinear), G.0226.06 (cooperative systems and optimization), G.0321.06 (Tensors), G.0302.07 (SVM/Kernel), research communities (ICCoS, ANMMM, MLDM); and IWT: PhD Grants, McKnow-E, Eureka-Flite; by the Belgian Federal Science Policy Office: IUAP P6/04 (DYSCO, Dynamical systems, control and optimization, 2007-2011) ; and by the EU: ERNSI.

References

[1] Willems, J.C.: From time series to linear system — Part I. Finite dimensional linear time invariant systems, Part II. Exact modelling, Part III. Approximate modelling. Automatica, **22**, 561–580 and 675–694 (1986), **23**, 87–115 (1987)
[2] Willems, J.C.: Paradigms and puzzles in the theory of dynamical systems. IEEE Transactions on Automatic Control, **36**, 259–294 (1991)
[3] Polderman, J.W., Willems, J.C.: Introduction to Mathematical Systems Theory: A Behavioral Approach. Springer-Verlag (1998)

[4] Willems, J.C.: Thoughts on system identification. In: Francis, B.A., Smith, M.C., Willems, J.C. (eds), Control of Uncertain Systems: Modelling, Approximation and Design, *Springer Verlag Lecture Notes on Control and Information Systems*, vol. 329, 389–416 (2006)

[5] Willems, J.C.: The behavioral approach to open and interconnected systems, Modeling by tearing, zooming, and linking. Control Systems Magazine, **27**, 49–99 (2007)

[6] Willems, J.C., Yamamoto, Y.: Behaviors defined by rational functions. Linear Algebra and Its Applications, **425**, 226–241 (2007)

Robust Controller Synthesis is Convex for Systems without Control Channel Uncertainties

Carsten W. Scherer

Abstract We consider an uncertain generalized plant whose control channel is not affected by uncertainties. It is shown that these configurations emerge in various concrete problems such as robust estimator or feed-forward controller design. Under the mere hypothesis that the (possibly non-linear) uncertainties are described by integral quadratic constraints, we reveal how one can translate robust controller synthesis to a problem of designing parametric and dynamic components in a standard plant configuration, and how this can be turned into a semi-definite program.

1 Introduction

It is well-known that the most general robust output-feedback controller synthesis problem cannot be easily solved by techniques from convex optimization. Still, in recent years, there has been a strong interest in classifying particularly structured robust synthesis problems that do have solutions in terms of semi-definite programming. Let us consider, for example, the intensively studied configuration as depicted in Figure 1 (see references in [20, 19]). Given the uncertain systems $G_1(\Delta)$, $G_2(\Delta)$

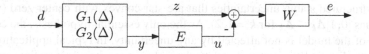

Fig. 1 Interconnection for Robust Estimator Synthesis

and a performance weight W, the problem is to design an estimator E such that

Carsten W. Scherer
Delft Center for Systems and Control, Delft University of Technology, Mekelweg 2, 2628 CD
Delft, The Netherlands, e-mail: c.w.scherer@tudelft.nl

P.M.J. Van den Hof et al. (eds.), *Model-Based Control: Bridging Rigorous Theory*
and Advanced Technology, DOI: 10.1007/978-1-4419-0895-7_2,
© Springer Science + Business Media, LLC 2009

the worst-case energy gain from the disturbance d to the weighted estimation error e is minimized. Clearly the interconnection in Figure 1 can be represented by the generalized plant configuration

$$\begin{pmatrix} e \\ y \end{pmatrix} = \begin{pmatrix} WG_1(\Delta) & -W \\ G_2(\Delta) & 0 \end{pmatrix} \begin{pmatrix} d \\ u \end{pmatrix}, \quad u = Ey$$

in which the control channel $u \to y$ is not affected by uncertainty.

Dually the interconnection for robust feed-forward control is shown in Figure 2 and relates to the generalized plant description

$$\begin{pmatrix} e \\ y \end{pmatrix} = \begin{pmatrix} G_1(\Delta)W & G_2(\Delta) \\ W & 0 \end{pmatrix} \begin{pmatrix} d \\ u \end{pmatrix}, \quad u = Ky$$

with a similar structural property (see references in [10, 12]).

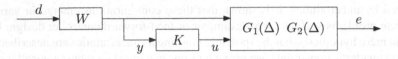

Fig. 2 Interconnection for Robust Feed-Forward Synthesis

Such interconnection structures emerge as well if considering synthesis problems with uncertain input- and output-performance weights as they are used in suppressing specific disturbances with time-varying frequencies or for describing coloring filters that are subject to parameter variations in order to capture disturbance spectrum uncertainty in stochastic control (see [8] and references therein).

The above sketched - and many more - problems can be subsumed to the following general formulation. Consider the uncertain generalized plant

$$\begin{pmatrix} e \\ y \end{pmatrix} = \begin{pmatrix} G_{11}(\Delta_1) & G_{12}(\Delta_2) \\ G_{21}(\Delta_1) & G_{22} \end{pmatrix} \begin{pmatrix} d \\ u \end{pmatrix} \quad \text{with } \Delta_1 \in \boldsymbol{\Delta}_1, \Delta_2 \in \boldsymbol{\Delta}_2 \quad (1)$$

where $d \to e$ is the performance channel, $u \to y$ is the control channel and $\boldsymbol{\Delta}_1, \boldsymbol{\Delta}_2$ are sets of structured stable uncertainties that are star-convex with center zero (which just means $[0,1]\boldsymbol{\Delta}_k \subset \boldsymbol{\Delta}_k$ for $k = 1,2$). As the only essential hypothesis, the control channel of the model is not affected by any uncertainty. In typical applications Δ_1 and Δ_2 are diagonally structured and might include time-invariant or time-varying rate-bounded parametric uncertainties, linear dynamic uncertainties or static and dynamic nonlinearities. Let us refer to Remark 1 for a brief discussion of variants of how the channels $u \to e$, $d \to e$ and $d \to y$ could be affected by uncertainties.

The goal is to design an output-feedback controller which robustly stabilizes (1) and which achieves a robust performance specification on the channel $d \to e$. In this paper both the uncertainties and the performance criterion are described by integral quadratic constraints (IQCs) with [14] being the main reference. If $\|.\|$ denotes the energy gain, this covers e.g. the pretty general robust model matching problem (see [4] and references therein) of designing a stable K which minimizes γ such that

$$\|G_{11}(\Delta_1) + G_{12}(\Delta_2)KG_{21}(\Delta_1)\| < \gamma \text{ for all } \Delta_k \in \Delta_k, \ k = 1, 2.$$

The paper is structured as follows. In Section 2 we discuss the construction of a suitable linear fractional representation of (1). Section 3 describes how to translate robust performance synthesis into a feasibility problem that involves parametric and dynamic decision variables, and Section 2 is devoted to turning this question into one of semi-definite programming. Further applications of this design framework are briefly sketched in Section 5, while a technical proof is deferred to the Appendix in Section 7.

Notation. We call a problem LMIable if it can be equivalently transformed into a linear matrix inequality (LMI) optimization problem. A set is LMIable if it can be represented as the feasible set of an LMI constraint [2, 3]. For complex matrices M we denote by M^* their conjugate transpose and use the abbreviation $\text{He}(M) = M + M^*$. We adopt the classical notions of interconnection stability as developed in [7]. In the sequel $\mathscr{L}_2$ denotes the space of (vector-valued) finite energy signals on $[0, \infty)$ while $\|d\|$ and $\hat{d}$ are the norm and the Fourier transform of $d \in \mathscr{L}_2$. We use the symbol $\mathbb{C}^0$ for the extended imaginary axis $i\mathbb{R} \cup \{\infty\}$. By abusing notation we do not distinguish between finite-dimensional linear time-invariant systems and their representation in terms of transfer matrices. For transfer matrices P and K the lower linear fractional transformation of P and K is denoted by $\star$ and defined as

$$\begin{pmatrix} P_{11} & P_{12} \\ P_{21} & P_{22} \end{pmatrix} \star K = P_{11} + P_{12}K(I - P_{22}K)^{-1}P_{21}$$

if $I - P_{22}K$ has a proper inverse. If Δ is nonlinear, we define $e = (\Delta \star P)(d)$ by

$$\begin{pmatrix} z \\ e \end{pmatrix} = \begin{pmatrix} P_{11} & P_{12} \\ P_{21} & P_{22} \end{pmatrix} \begin{pmatrix} w \\ d \end{pmatrix}, \ w = \Delta(z)$$

if $I - P_{11}\Delta$ has a causal inverse; note that $e = P_{22}d + P_{21}\Delta((I - P_{11}\Delta)^{-1}(P_{21}d))$. Finally $P = \left[\begin{array}{c|c} A & B \\ \hline C & D \end{array} \right]$ means $P(s) = C(sI - A)^{-1}B + D$.

2 System Interconnections and Performance Specification

We assume that (1) admits a linear fractional representation. More precisely let us suppose that

$$\begin{pmatrix} G_{11}(\Delta_1) \\ G_{21}(\Delta_1) \end{pmatrix} = \Delta_1 \star \begin{pmatrix} P_{11} & P_{13} \\ P_{31} & P_{33} \\ P_{43} & P_{44} \end{pmatrix} \text{ and } G_{12}(\Delta_2) = \Delta_2 \star \begin{pmatrix} P_{22} & P_{24} \\ P_{32} & P_{34} \end{pmatrix}$$

where we assume that

$$I - P_{11}\Delta_1 \text{ and } I - P_{22}\Delta_2 \text{ have a causal inverse for all } \Delta_1 \in \mathbf{\Delta}_1, \Delta_2 \in \mathbf{\Delta}_2. \quad (2)$$

With $P_{44} = G_{22}$ this leads to the following generalized plant description of (1):

$$\begin{pmatrix} z_1 \\ z_2 \\ e \\ y \end{pmatrix} = \underbrace{\left(\begin{array}{cc|cc} P_{11} & 0 & P_{13} & 0 \\ 0 & P_{22} & 0 & P_{24} \\ \hline P_{31} & P_{32} & P_{33} & P_{34} \\ P_{41} & 0 & P_{43} & P_{44} \end{array} \right)}_{P} \begin{pmatrix} w_1 \\ w_2 \\ d \\ u \end{pmatrix}, \quad w_1 = \Delta_1(z_1), \quad w_2 = \Delta_2(z_2). \quad (3)$$

As well-known, the closed-loop interconnection of (1) with a dynamic controller $u = Ky$ can be obtained by first computing $P \star K$ and then interconnecting the uncertainties. Note that

$$P \star K = \underbrace{\left(\begin{array}{cc|c} P_{11} & 0 & P_{13} \\ P_{24}K(I - P_{44}K)^{-1}P_{41} & P_{22} & P_{24}K(I - P_{44}K)^{-1}P_{43} \\ \hline P_{31} + P_{34}K(I - P_{44}K)^{-1}P_{41} & P_{32} & P_{33} + P_{34}K(I - P_{44}K)^{-1}P_{43} \end{array} \right)}_{\mathscr{P}}.$$

Hence the controlled uncertain system admits the description

$$\begin{pmatrix} z_1 \\ z_2 \\ e \end{pmatrix} = \underbrace{\left(\begin{array}{cc|c} P_{11} & 0 & P_{13} \\ \mathscr{P}_{21} & P_{22} & \mathscr{P}_{23} \\ \hline \mathscr{P}_{31} & P_{32} & \mathscr{P}_{33} \end{array} \right)}_{\mathscr{P}} \begin{pmatrix} w_1 \\ w_2 \\ d \end{pmatrix}, \quad w_1 = \Delta_1(z_1), \quad w_2 = \Delta_2(z_2). \quad (4)$$

Here we have indicated those blocks which explicitly depend on the controller by calligraphic symbols. The particular structure how the controller enters the individual blocks of $\mathscr{P}$ will be of crucial relevance in the sequel.

We assume that there exists some K which (internally) stabilizes P. Clearly P_{11}, P_{13}, P_{22}, P_{32} must hence be stable. If a nominally stabilizing controller K also achieves robust stability, then $d \in \mathscr{L}_2$ implies for the controlled uncertain system (4) that $e \in \mathscr{L}_2$. With some real symmetric matrix

$$\Pi_p = \begin{pmatrix} Q_p & S_p \\ S_p^T & T_p^T T_p \end{pmatrix} \quad \text{where } T_p \text{ has full column rank,} \quad (5)$$

the desired quadratic performance specification is then assumed to be expressed as

$$\int_{-\infty}^{\infty} \begin{pmatrix} \hat{d}(i\omega) \\ \hat{e}(i\omega) \end{pmatrix}^* \Pi_p \begin{pmatrix} \hat{d}(i\omega) \\ \hat{e}(i\omega) \end{pmatrix} d\omega \leq -\varepsilon \|d\|_2^2 \quad \text{for all } d \in \mathscr{L}_2 \quad (6)$$

and some $\varepsilon > 0$. Robust quadratic performance is achieved if this inequality holds for all trajectories of the uncertain system (4). Let us recall the special case $Q_p = -I$, $S_p = 0$, $T_p = I$; then (6) means that the $\mathscr{L}_2$-gain of $d \to e$ is bounded by one.

Remark 1. We might be confronted with an uncertain model

$$\begin{pmatrix} e \\ y \end{pmatrix} = \begin{pmatrix} G_{11}(\Delta_1) & G_{12}(\Delta_3) \\ G_{21}(\Delta_2) & G_{22} \end{pmatrix} \begin{pmatrix} d \\ u \end{pmatrix} \tag{7}$$

whose blocks are affected by uncertainties Δ_1, Δ_2 and Δ_3 in the different classes $\boldsymbol{\Delta}_1$, $\boldsymbol{\Delta}_2$ and $\boldsymbol{\Delta}_3$. One would then construct the linear fractional representation

$$\begin{pmatrix} z_1 \\ z_2 \\ z_3 \\ \hline e \\ y \end{pmatrix} = \left(\begin{array}{ccc|cc} P_{11} & 0 & 0 & P_{14} & 0 \\ 0 & P_{22} & 0 & P_{24} & 0 \\ 0 & 0 & P_{33} & 0 & P_{35} \\ \hline P_{41} & 0 & P_{43} & P_{44} & P_{45} \\ 0 & P_{52} & 0 & P_{54} & G_{22} \end{array} \right) \begin{pmatrix} w_1 \\ w_2 \\ w_3 \\ \hline d \\ u \end{pmatrix}, \quad \begin{pmatrix} w_1 \\ w_2 \\ w_3 \end{pmatrix} = \begin{pmatrix} \Delta_1(z_1) \\ \Delta_2(z_2) \\ \Delta_3(z_3) \end{pmatrix}.$$

This can be obviously subsumed to (3) and is, therefore, just a special version of the generalized plant considered throughout this paper.

Moreover, all our results could as well be based on a dual construction; one would then combine a linear fractional representation of $e = G_{11}(\Delta_1)(d) + G_{12}(\Delta_1)(u)$ with one of $y = G_{21}(\Delta_2)(d)$.

3 Robust Performance Analysis

For $k = 1, 2$ let $\boldsymbol{\Pi}_k$ denote a class of valid IQC for $\boldsymbol{\Delta}_k$ [14]. This just means that any $\Pi_k \in \boldsymbol{\Pi}_k$, which is said to be a multiplier, is a transfer matrix that is bounded and Hermitian-valued on the imaginary axis and for which we have

$$\int_{-\infty}^{\infty} \begin{pmatrix} \widehat{\Delta_k(w)}(i\omega) \\ \hat{w}(i\omega) \end{pmatrix}^* \Pi_k(i\omega) \begin{pmatrix} \widehat{\Delta_k(w)}(i\omega) \\ \hat{w}(i\omega) \end{pmatrix} d\omega \geq 0 \quad \text{for all } w \in \mathscr{L}_2 \tag{8}$$

and for all $\Delta_k \in \boldsymbol{\Delta}_k$, $k = 1, 2$. In the sequel we sometimes introduce the partition $\Pi_k = \begin{pmatrix} Q_k & S_k \\ S_k^* & R_k \end{pmatrix}$ according to the dimensions of the signals $\Delta_k(w)$ and w in (8). Since $\boldsymbol{\Delta}_k$ is star-convex it contains 0; therefore (8) implies $R_k \succcurlyeq 0$ on $\mathbb{C}^0$ for $k = 1, 2$.

If K stabilizes P (implying stability of $\mathscr{P}$), it achieves robust stability and robust performance in case there exists some $\Pi_1 \in \boldsymbol{\Pi}_1$ and $\Pi_2 \in \boldsymbol{\Pi}_2$ such that the following frequency domain inequality (FDI) is satisfied:

$$\begin{pmatrix} I & 0 & 0 \\ P_{11} & 0 & P_{13} \end{pmatrix}^* \Pi_1 \begin{pmatrix} I & 0 & 0 \\ P_{11} & 0 & P_{13} \end{pmatrix} + \begin{pmatrix} 0 & I & 0 \\ \mathscr{P}_{21} & P_{22} & \mathscr{P}_{23} \end{pmatrix}^* \Pi_2 \begin{pmatrix} 0 & I & 0 \\ \mathscr{P}_{21} & P_{22} & \mathscr{P}_{23} \end{pmatrix} +$$

$$+ \begin{pmatrix} 0 & 0 & I \\ \mathscr{P}_{31} & P_{32} & \mathscr{P}_{33} \end{pmatrix}^* \Pi_p \begin{pmatrix} 0 & 0 & I \\ \mathscr{P}_{31} & P_{32} & \mathscr{P}_{33} \end{pmatrix} \prec 0 \quad \text{on } \mathbb{C}^0. \tag{9}$$

Indeed, since the right-lower blocks of Π_1, Π_2 and Π_p are positive-semi-definite on $\mathbb{C}^0$, we infer from (9) that

$$\begin{pmatrix} I \\ P_{11} \end{pmatrix}^* \Pi_1 \begin{pmatrix} I \\ P_{11} \end{pmatrix} \prec 0 \quad \text{and} \quad \begin{pmatrix} I \\ P_{22} \end{pmatrix}^* \Pi_2 \begin{pmatrix} I \\ P_{22} \end{pmatrix} \prec 0 \quad \text{on } \mathbb{C}^0. \qquad (10)$$

The standard IQC theorem [14] hence implies that the inverses of $I - P_{11}\Delta_1$ and $I - P_{22}\Delta_2$ are stable. In turn this guarantees robust stability of (4). If $\varepsilon > 0$ is sufficiently small, (9) persists to hold if replacing the right-hand side by $-\frac{\varepsilon}{2\pi}I$. If we then choose any trajectory of (4) with $d \in \mathscr{L}_2$, we infer that all other signals are in $\mathscr{L}_2$ as well. Evaluation of the quadratic form (9) for $\text{col}(\widehat{w}_1(i\omega), \widehat{w}_2(i\omega), \widehat{d}(i\omega))$ leads, with the system description (4), to

$$\sum_{k=1}^{2} \begin{pmatrix} \widehat{w}_k(i\omega) \\ \widehat{z}_k(i\omega) \end{pmatrix}^* \Pi_k(i\omega) \begin{pmatrix} \widehat{w}_k(i\omega) \\ \widehat{z}_k(i\omega) \end{pmatrix} + \begin{pmatrix} \widehat{d}(i\omega) \\ \widehat{e}(i\omega) \end{pmatrix}^* \Pi_p \begin{pmatrix} \widehat{d}(i\omega) \\ \widehat{e}(i\omega) \end{pmatrix} \leq -\frac{\varepsilon}{2\pi} \|\widehat{d}(i\omega)\|^2$$

for all $\omega \in \mathbb{R} \cup \{\infty\}$. Integration and exploiting (8) implies the validity of (6).

We have sketched the standard arguments which reduce the desired robustness specification on the closed-loop system to an FDI. For fixed mulitpliers Π_1 and Π_2, the search for a stabilizing controller which renders (9) satisfied is well-know to be LMIable. However, for reduced conservatism, one wishes to view Π_1 and Π_2 together with the controller as optimization variables. Even if considering "nicely" (conicly) parameterized classes of multipliers $\boldsymbol{\Pi}_k$, all standard procedures which are convexifying for fixed multipliers lead to bi-linear matrix inequalities if the multipliers are also considered as variables.

As one of the key contributions of this paper we show that one can equivalently rewrite the FDI (9) into one in Π_1 and Π_2^{-1} which requires all elements of $\boldsymbol{\Pi}_2$ to be non-singular. This is formulated in the following auxiliary result whose proof is found in the Appendix (Section 7).

Lemma 1. *Suppose that any $\Pi_2 \in \boldsymbol{\Pi}_2$ is non-singular and that the left-upper block of its inverse is negative semi-definite on $\mathbb{C}^0$. Then (9) is equivalent to*

$$\begin{pmatrix} 0 & 0 & 0 & 0 \\ 0 & Q_p & 0 & 0 \\ 0 & 0 & 0 & 0 \\ 0 & 0 & 0 & -I \end{pmatrix} + \begin{pmatrix} I & 0 \\ 0 & I \\ 0 & 0 \\ 0 & 0 \end{pmatrix} \begin{pmatrix} I & 0 \\ P_{11} & P_{13} \end{pmatrix}^* \Pi_1 \begin{pmatrix} I & 0 \\ P_{11} & P_{13} \end{pmatrix} \begin{pmatrix} I & 0 \\ 0 & I \\ 0 & 0 \\ 0 & 0 \end{pmatrix}^T -$$

$$- \begin{pmatrix} 0 & 0 \\ 0 & S_p \\ I & 0 \\ 0 & T_p \end{pmatrix} \begin{pmatrix} P_{22} & -I \\ P_{32} & 0 \end{pmatrix} \Pi_2^{-1} \begin{pmatrix} P_{22} & -I \\ P_{32} & 0 \end{pmatrix}^* \begin{pmatrix} 0 & 0 \\ 0 & S_p \\ I & 0 \\ 0 & T_p \end{pmatrix}^T +$$

$$+ He \begin{pmatrix} 0 & 0 \\ 0 & S_p \\ I & 0 \\ 0 & T_p \end{pmatrix} \begin{pmatrix} \mathscr{P}_{21} & \mathscr{P}_{23} \\ \mathscr{P}_{31} & \mathscr{P}_{33} \end{pmatrix} \begin{pmatrix} I & 0 & 0 & 0 \\ 0 & I & 0 & 0 \end{pmatrix} \prec 0 \ on \ \mathbb{C}^0. \quad (11)$$

In view of the IQC dualization techniques in [12], this novel partial dualization of the robust performance FDI leads to an additive separation of the transfer matrices into a part that depends on the multipliers Π_1 and Π_2^{-1} and another part which is only affected by the controller. This is the essential feature in order to convexify the considered synthesis problem. For actual computations we assume that the multipliers Π_1 and Π_2^{-1} are parameterized as

$$\Pi_1 = \{\Psi_1^* M_1 \Psi_1 : M_1 \in \mathcal{M}_1\} \text{ and } \Pi_2^{-1} = \{\Psi_2^* M_2 \Psi_2 : M_2 \in \mathcal{M}_2\}$$

with LMIable sets $\mathcal{M}_1$ and $\mathcal{M}_2$ of real symmetric matrices and with some fixed transfer matrices Ψ_1 and Ψ_2 that have no poles on the extended imaginary axis $\mathbb{C}^0$. It is stressed that we do not require stability of these outer factors in the sequel. Note that the parametrization of Π_1 naturally complies with many multiplier classes that have been suggested in the literature. Since this is not generically true for Π_2^{-1}, we refer to [12] for a more detailed discussion of this issue. Recall that P_{11}, P_{13} and P_{22}, P_{32} are stable. Therefore one can describe the family obtained by the sum of the first three matrices in (11) as

$$\Psi^* M \Psi \text{ with parameter } M \in \mathcal{M}$$

where $\mathcal{M}$ is an LMIable set of real symmetric structured matrices and Ψ is a tall transfer matrix without poles in $\mathbb{C}^0$. Let us finally observe that the last term in (11) just equals $\text{He}(H \star K)$ if defining

$$H := \begin{pmatrix} 0 & 0 & 0 \\ 0 & S_p & 0 \\ I & 0 & 0 \\ 0 & T_p & 0 \\ \hline 0 & 0 & I \end{pmatrix} \begin{pmatrix} 0 & 0 & P_{24} \\ P_{31} & P_{33} & P_{34} \\ P_{41} & P_{43} & P_{44} \end{pmatrix} \begin{pmatrix} I & 0 & 0 & 0 & 0 \\ 0 & I & 0 & 0 & 0 \\ 0 & 0 & 0 & 0 & I \end{pmatrix}.$$

This reveals how the considered robust performance synthesis problem subsumes to the generic optimization problem as considered in the subsequent Section 2.

4 Parametric-Dynamic Feasibility Problems

The previous section revealed that a surprisingly large variety of robust synthesis questions can be reduced to the following problem: Let us be given a generalized plant H, a transfer matrix Ψ without poles on the imaginary axis and any LMIable set $\mathcal{M}$ of symmetric matrices of dimension compatible with the number of rows of Ψ. Determine some matrix $M \in \mathcal{M}$ and a controller K which stabilizes H such that

$$\Psi^* M \Psi + (H \star K) + (H \star K)^* \prec 0 \text{ on } \mathbb{C}^0. \tag{12}$$

We choose to call this a parametric-dynamic feasibility problem since it involves the common search for the parameter $M \in \mathcal{M}$ and the dynamic controller K in order to render (12) satisfied. Observe that the FDI can as well be expressed as

$$\begin{pmatrix} I \\ \Psi \\ H \star K \end{pmatrix}^* \underbrace{\begin{pmatrix} 0 & 0 & I \\ 0 & M & 0 \\ I & 0 & 0 \end{pmatrix}}_{M_e} \begin{pmatrix} I \\ \Psi \\ H \star K \end{pmatrix} \prec 0 \text{ on } \mathbb{C}^0. \tag{13}$$

Despite this relatively simplistic and specific formulation, we will briefly illustrate in Section 5 that it also straightforwardly covers the problem classes in [5, 6, 11, 15].

In the sequel we use the minimal state-space realizations

$$\Psi = \left[\begin{array}{c|c} A_\Psi & B_\Psi \\ \hline C_\Psi & D_\Psi \end{array} \right] \text{ and } H = \begin{pmatrix} H_{11} & H_{12} \\ H_{21} & H_{22} \end{pmatrix} = \left[\begin{array}{c|cc} A_H & B_{H1} & B_H \\ \hline C_{H1} & D_{H1} & E_H \\ C_H & F_H & 0 \end{array} \right],$$

where the partition of H is induced by the dimension of K. Note that A_Ψ has no eigenvalues on the imaginary axis. Moreover (A_H, B_H) and (A_H, C_H) are assumed to be stabilizable and detectable such that the existence of a stabilizing controller for H is guaranteed. Realizations of controllers K are denoted as

$$K = \left[\begin{array}{c|c} A_K & B_K \\ \hline C_K & D_K \end{array} \right].$$

Then the closed-loop transfer matrix admits the natural state-space description

$$H \star K = H_{cl} = \left[\begin{array}{c|c} A_{cl} & B_{cl} \\ \hline C_{cl} & D_{cl} \end{array} \right] = \left[\begin{array}{cc|c} A_H + B_H D_K C_H & B_H C_K & B_{H1} + B_H D_K F_H \\ C_H B_K & A_K & B_K F_H \\ \hline C_{H1} + E_H D_K C_H & E_H C_K & D_{H1} + E_H D_K F_H \end{array} \right].$$

This leads to the following realization of the dynamics in the outer factor of (13):

$$\begin{pmatrix} \Psi \\ H \star K \end{pmatrix} = \begin{pmatrix} \Psi \\ H_{cl} \end{pmatrix} = \left[\begin{array}{cc|c} A_\Psi & 0 & B_\Psi \\ 0 & A_{cl} & B_{cl} \\ \hline C_\Psi & 0 & D_\Psi \\ 0 & C_{cl} & D_{cl} \end{array} \right].$$

The very same transfer matrix is obtained by interconnecting K with

$$\begin{pmatrix} \Psi & 0 \\ H_{11} & H_{12} \\ H_{21} & H_{22} \end{pmatrix} = \left[\begin{array}{cc|cc} A_\Psi & 0 & B_\Psi & 0 \\ 0 & A_H & B_{H1} & B_{H2} \\ \hline C_\Psi & 0 & D_\Psi & 0 \\ 0 & C_{H1} & D_{H11} & D_{H12} \\ 0 & C_{H2} & D_{H21} & 0 \end{array} \right] =: \left[\begin{array}{c|cc} A & B_1 & B \\ \hline C_1 & D_1 & E \\ C & F & 0 \end{array} \right] \tag{14}$$

by a lower linear fractional transformation as

$$
\begin{pmatrix} \Psi \\ H_{cl} \end{pmatrix} = \begin{pmatrix} \Psi & 0 \\ H_{11} & H_{12} \\ H_{21} & H_{22} \end{pmatrix} \star K = \left[\begin{array}{cc|c} A+BD_KC & BC_K & B_1+BD_KF \\ CB_K & A_K & B_KF \\ \hline C_1+ED_KC & EC_K & D_1+ED_KF \end{array} \right]. \tag{15}
$$

Let us finally introduce a shorthand notation for the closed-loop realization matrices in terms of calligraphic symbols and observe that they can be written in the following three different ways:

$$
\left(\begin{array}{c|c} \mathscr{A} & \mathscr{B} \\ \hline \mathscr{C} & \mathscr{D} \end{array} \right) := \left(\begin{array}{cc|c} A_\Psi & 0 & B_\Psi \\ 0 & A_{cl} & B_{cl} \\ \hline C_\Psi & 0 & D_\Psi \\ 0 & C_{cl} & D_{cl} \end{array} \right) = \left(\begin{array}{cc|c} A+BD_KC & BC_K & B_1+BD_KF \\ B_KC & A_K & B_KF \\ \hline C_1+ED_KC & EC_K & D_1+ED_KF \end{array} \right) =
$$

$$
= \left(\begin{array}{cc|c} A & 0 & B_1 \\ 0 & 0 & 0 \\ \hline C_1 & 0 & D_1 \end{array} \right) + \left(\begin{array}{cc} B & 0 \\ 0 & I \\ \hline E & 0 \end{array} \right) \begin{pmatrix} D_K & C_K \\ B_K & A_K \end{pmatrix} \left(\begin{array}{cc|c} C & 0 & F \\ 0 & I & 0 \end{array} \right). \tag{16}
$$

4.1 Analysis

Let us now suppose that K stabilizes H and leads to (13). This means that A_{cl} is Hurwitz and that

$$
\left[\begin{array}{c|c} \mathscr{A} & \mathscr{B} \\ \hline 0 & I \\ \mathscr{C} & \mathscr{D} \end{array} \right]^* M_e \left[\begin{array}{c|c} \mathscr{A} & \mathscr{B} \\ \hline 0 & I \\ \mathscr{C} & \mathscr{D} \end{array} \right] \prec 0 \quad \text{on } \mathbb{C}^0. \tag{17}
$$

Since A_Ψ does not have eigenvalues on the imaginary axis, the same holds for $\mathscr{A}$. We can apply the Kalman Yakubovich Popov lemma [1] to (17) in order to infer that its validity is equivalent to the existence of a symmetric solution $\mathscr{X}$ of the LMI

$$
\begin{pmatrix} \mathscr{A}^T\mathscr{X} + \mathscr{X}\mathscr{A} + \mathscr{C}^T \begin{pmatrix} M & 0 \\ 0 & 0 \end{pmatrix} \mathscr{C} & \mathscr{X}\mathscr{B} + \mathscr{C}^T \begin{pmatrix} MD_\Psi \\ I \end{pmatrix} \\ \mathscr{B}^T\mathscr{X} + \begin{pmatrix} D_\Psi^T M & I \end{pmatrix} \mathscr{C} & \begin{pmatrix} 0 & I \end{pmatrix}\mathscr{D} + \mathscr{D}^T \begin{pmatrix} 0 \\ I \end{pmatrix} + D_\Psi^T M D_\Psi \end{pmatrix} \prec 0.
\tag{18}
$$

In view of (16) let us note that the left-upper block of this inequality reads as

$$
\begin{pmatrix} A_\Psi & 0 \\ 0 & A_{cl} \end{pmatrix}^T \begin{pmatrix} \mathscr{X}_{11} & \mathscr{X}_{12} \\ \mathscr{X}_{21} & \mathscr{X}_{22} \end{pmatrix} + \begin{pmatrix} \mathscr{X}_{11} & \mathscr{X}_{12} \\ \mathscr{X}_{21} & \mathscr{X}_{22} \end{pmatrix} \begin{pmatrix} A_\Psi & 0 \\ 0 & A_{cl} \end{pmatrix} + \begin{pmatrix} C_\Psi^T M C_\Psi & 0 \\ 0 & 0 \end{pmatrix} \prec 0
$$

with a partition of $\mathscr{X}$ compatible with that of $\mathrm{diag}(A_\Psi, A_{cl})$. Since M is indefinite, stability of A_{cl} does in general *not* translate into the simple constraint $\mathscr{X} \succ 0$ on the solution of (18). However, we observe that (18) always implies

$$
A_{cl}^T \mathscr{X}_{22} + \mathscr{X}_{22} A_{cl} \prec 0. \tag{19}
$$

Since A_{cl} is Hurwitz we infer $\mathscr{X}_{22} \succ 0$. If we define

$$\mathscr{J} = \begin{pmatrix} I \\ 0 \end{pmatrix} \text{ in the partition of } \mathscr{A} = \begin{pmatrix} A_\Psi & 0 \\ 0 & A_{cl} \end{pmatrix},$$

it is easy to verify that $\mathscr{X}_{22} \succ 0$ is equivalent to the existence of some $\hat{M} = \hat{M}^T$ with

$$\mathscr{X} + \mathscr{J}\hat{M}\mathscr{J}^T \succ 0. \tag{20}$$

Conversely, let there exist $\mathscr{X}$, $\hat{M}$ with (18) and (20). Then $\mathscr{X}_{22} \succ 0$ and the consequence (19) of (18) reveals that A_{cl} is Hurwitz; hence K stabilizes H. Moreover, (18) leads to the validity of the FDI (13). This proves the following intermediate result.

Lemma 2. *The controller K stabilizes H and renders (13) satisfied iff there exist solutions $\mathscr{X}$ and $\hat{M}$ of the inequalities (18) and (20).*

Most parts of this result are standard. The key twist is the very elementary characterization of the fact that K stabilizes H by the inequality (20). This formulation of the analysis result is tailored towards a convenient proof for synthesis.

4.2 Synthesis

On the basis of Lemma 2 let us now develop the required transformation of $\mathscr{X}$ and the state-space description of K which convexifies both (18) and (20).

For this purpose we assume that there exists a controller K for which one can find M, $\hat{M}$, $\mathscr{X}$ that satisfy (18) and (20). According to the sizes of A and A_K in (16) let us partition $\mathscr{X}$ as

$$\mathscr{X} = \begin{pmatrix} X & U \\ U^T & \hat{X} \end{pmatrix}.$$

Note that the dimension of A_K can be taken at least as large as that of A (by adding uncontrollable/uobservalbe modes in the controller if necessary). Hence U is a wide matrix. By perturbation (if necessary) we can then assume without loss of generality that $\mathscr{X}$ and $\hat{X}$ are non-singular and that U has full row rank. This implies that $X - U\hat{X}^{-1}U^T$ is invertible. Let us now partition this latter matrix as A in (14). An additional perturbation allows to make sure that the right-lower block of $X - U\hat{X}^{-1}U^T$ is non-singular as well. Hence this block can be written as W_{22}^{-1} for some non-singular real symmetric matrix W_{22}, and we can then sequentially determine real matrices W_{21} and $W_{11} = W_{11}^T$ with

$$X - U\hat{X}^{-1}U^T = \begin{pmatrix} W_{11} + W_{21}^T W_{22}^{-1} W_{21} & W_{21}^T W_{22}^{-1} \\ W_{22}^{-1} W_{21} & W_{22}^{-1} \end{pmatrix}. \tag{21}$$

Let us collect the blocks W_{11}, W_{21}, W_{22} into W and introduce the structured matrices Y and Z as follows:

$$W := \begin{pmatrix} W_{11} & W_{21}^T \\ W_{21} & W_{22} \end{pmatrix}, \quad Y := \begin{pmatrix} I & -W_{21}^T \\ 0 & W_{22} \end{pmatrix}, \quad Z := \begin{pmatrix} W_{11} & 0 \\ W_{21} & I \end{pmatrix}. \tag{22}$$

Due to (21) one easily checks that $Z = Y(X - U\hat{X}^{-1}U^T)$ and hence

$$Z(X - U\hat{X}^{-1}U^T)^{-1} = Y. \tag{23}$$

Note that both Y and Z are non-singular. Then the matrix

$$\mathscr{Y} = \begin{pmatrix} Z & 0 \\ X & U \end{pmatrix} \mathscr{X}^{-1}$$

has full row rank. Moreover, just because $(\,X\ U\,)$ is the upper block row of $\mathscr{X}$, the lower block row of $\mathscr{Y}$ must equal $(\,I\ 0\,)$. Since $(X - U\hat{X}^{-1}U^T)^{-1}$ is the left-upper block of $\mathscr{X}^{-1}$, (23) implies that the left-upper block of $\mathscr{Y}$ must equal Y. Therefore we arrive at the following structure of $\mathscr{Y}$ together with a crucial factorization of $\mathscr{X}$:

$$\mathscr{Y} = \begin{pmatrix} Y & V \\ I & 0 \end{pmatrix} \quad \text{and} \quad \mathscr{Y}\mathscr{X} = \begin{pmatrix} Z & 0 \\ X & U \end{pmatrix}. \tag{24}$$

Since $\mathscr{Y}$ has full row rank, (18) implies that

$$\begin{pmatrix} \mathrm{He}(\mathscr{Y}\mathscr{X}\mathscr{A}\mathscr{Y}^T) + \mathscr{Y}\mathscr{C}^T \begin{pmatrix} M & 0 \\ 0 & 0 \end{pmatrix} \mathscr{C}\mathscr{Y}^T & \mathscr{Y}\mathscr{X}\mathscr{B} + \mathscr{Y}\mathscr{C}^T \begin{pmatrix} MD_\Psi \\ I \end{pmatrix} \\ \mathscr{B}^T\mathscr{X}\mathscr{Y}^T + (\,D_\Psi^T M\ I\,)\mathscr{C}\mathscr{Y}^T & \mathrm{He}((\,0\ I\,)\mathscr{D}) + D_\Psi^T M D_\Psi \end{pmatrix} \prec 0. \tag{25}$$

With the last relation in (16) and with (24) we obtain

$$\left(\begin{array}{c|c} \mathscr{Y}\mathscr{X}\mathscr{A}\mathscr{Y}^T & \mathscr{Y}\mathscr{X}\mathscr{B} \\ \hline \mathscr{C}\mathscr{Y}^T & \mathscr{D} \end{array} \right) = \left(\begin{array}{c|c} \mathscr{Y}\mathscr{X} & 0 \\ \hline 0 & I \end{array} \right) \left(\begin{array}{c|c} \mathscr{A} & \mathscr{B} \\ \hline \mathscr{C} & \mathscr{D} \end{array} \right) \left(\begin{array}{c|c} \mathscr{Y}^T & 0 \\ \hline 0 & I \end{array} \right) =$$

$$= \left(\begin{array}{cc|c} ZAY^T & ZA & ZB_1 \\ XAY^T & XA & XB_1 \\ \hline C_1Y^T & C_1 & D_1 \end{array} \right) + \left(\begin{array}{cc} ZB & 0 \\ XB & U \\ \hline E & 0 \end{array} \right) \left(\begin{array}{cc} D_K & C_K \\ B_K & A_K \end{array} \right) \left(\begin{array}{cc|c} CY^T & C & F \\ V^T & 0 & 0 \end{array} \right) =$$

$$= \left(\begin{array}{cc|c} ZAY^T & ZA & ZB_1 \\ 0 & XA & XB_1 \\ \hline C_1Y^T & C_1 & D_1 \end{array} \right) + \left(\begin{array}{cc} ZB & 0 \\ 0 & I \\ \hline E & 0 \end{array} \right) N \left(\begin{array}{cc|c} I & 0 & 0 \\ 0 & C & F \end{array} \right)$$

in case we introduce the new variable

$$N = \begin{pmatrix} 0 & 0 \\ XAY^T & 0 \end{pmatrix} + \begin{pmatrix} I & 0 \\ XB & U \end{pmatrix} \begin{pmatrix} D_K & C_K \\ B_K & A_K \end{pmatrix} \begin{pmatrix} CY^T & I \\ V^T & 0 \end{pmatrix}. \tag{26}$$

This motivates the definition

$$\begin{pmatrix} A(N,W,X) & B(N,W,X) \\ C(N,W) & D(N) \end{pmatrix} := \left(\begin{array}{cc|c} ZAY^T & ZA & ZB_1 \\ 0 & XA & XB_1 \\ \hline C_1Y^T & C_1 & D_1 \end{array} \right) + \begin{pmatrix} ZB & 0 \\ 0 & I \\ E & 0 \end{pmatrix} N \left(\begin{array}{cc|c} I & 0 & 0 \\ 0 & C & F \end{array} \right)$$

with which (25) just reads as

$$\begin{pmatrix} \mathrm{He}(A(N,W,X)) + C(N,W)^T \begin{pmatrix} M & 0 \\ 0 & 0 \end{pmatrix} C(N,W) & * \\ B(N,W,X)^T + (D_\Psi^T M \; I) C(N,W) & \mathrm{He}\left((0\;I) D(N) \right) + D_\Psi^T M D_\Psi \end{pmatrix} \prec 0.$$
(27)

If we zoom in and exploit the special structure of the matrices in (14) it is not hard to see that this is actually an LMI in M and (N,W,X). Indeed with (22) we have

$$\begin{pmatrix} ZAY^T & ZB \\ C_1Y^T & 0 \end{pmatrix} = \begin{pmatrix} \begin{pmatrix} W_{11}A_\Psi & 0 \\ W_{21}A_\Psi - A_H W_{21} & A_H W_{22} \end{pmatrix} & \begin{pmatrix} 0 \\ B_H \end{pmatrix} \\ \begin{pmatrix} C_\Psi & 0 \\ -C_{H1}W_{21} & C_{H1}W_{22} \end{pmatrix} & 0 \end{pmatrix}$$

which reveals *affine dependence* of A, B, C and D on (N,W,X). Moreover, again with (14) we observe that

$$\begin{pmatrix} M & 0 \\ 0 & 0 \end{pmatrix} (E\;\;0) N \begin{pmatrix} I & 0 \\ 0 & C \end{pmatrix} = \begin{pmatrix} M & 0 \\ 0 & 0 \end{pmatrix} \begin{pmatrix} 0 & 0 \\ D_{H12} & 0 \end{pmatrix} N \begin{pmatrix} I & 0 \\ 0 & C \end{pmatrix} = 0$$

and hence

$$C(N,W)^T \begin{pmatrix} M & 0 \\ 0 & 0 \end{pmatrix} C(N,W) = \begin{pmatrix} C_\Psi^T M C_\Psi & 0 & C_\Psi^T M C_\Psi & 0 \\ 0 & 0 & 0 & 0 \\ C_\Psi^T M C_\Psi & 0 & C_\Psi^T M C_\Psi & 0 \\ 0 & 0 & 0 & 0 \end{pmatrix}$$

is actually *independent from* (N,W) and, clearly, still affine in M. Similarly

$$(D_\Psi^T M \; I) C(N,W) =$$
$$= (D_\Psi^T M C_\Psi - C_{H1}W_{21} \quad C_{H1}W_{22} \quad D_\Psi^T M C_\Psi \quad C_{H1}) + (D_{H12} \; 0) W \begin{pmatrix} I & 0 \\ 0 & C \end{pmatrix}$$
(28)

is affine in M and (N,W). All this shows that the left-hand side of (27) depends affinely on M and (N,W,X) as claimed.

Let us now consider (20). Again since $\mathscr{Y}$ has full row rank this inequality implies

$$\mathscr{Y} \mathscr{X} \mathscr{Y}^T + \mathscr{Y} \mathscr{J} \hat{M} (\mathscr{Y} \mathscr{J})^T \succ 0.$$
(29)

It is easy to check that (29) reads explicitly as

$$\left(\begin{pmatrix} W_{11}+\hat{M} & 0 \\ 0 & W_{22} \end{pmatrix} \quad \begin{pmatrix} W_{11}+\hat{M} & 0 \\ W_{21} & I \end{pmatrix} \right.$$
$$\left. \begin{pmatrix} W_{11}+\hat{M} & W_{21}^T \\ 0 & I \end{pmatrix} \quad \begin{pmatrix} \hat{M} & 0 \\ 0 & 0 \end{pmatrix}+X \right) \succ 0. \tag{30}$$

By taking the Schur complement, (30) is equivalent to $W_{11}+\hat{M} \succ 0$ and

$$\begin{pmatrix} W_{22} & \begin{pmatrix} W_{21} & I \end{pmatrix} \\ \begin{pmatrix} W_{21}^T \\ I \end{pmatrix} & X-\begin{pmatrix} W_{11} & 0 \\ 0 & 0 \end{pmatrix} \end{pmatrix} \succ 0 \tag{31}$$

which is clearly an LMI constraint on (W,X). All this proves necessity in the following main synthesis result of this paper.

Theorem 1. *There exists a controller K which stabilizes H and for which there are $M \in \mathcal{M}$ and symmetric $\mathcal{X}$, $\hat{M}$ with (18) and (20) if and only if the LMIs (27) and (31) admit a solution $M \in \mathcal{M}$ and (N,W,X).*

To conclude the proof let us suppose that $M \in \mathcal{M}$ and (N,W,X) satisfy (27) and (31). Define Y and Z according to (22). Taking the Schur complement shows that (31) implies $W_{22} \succ 0$ and

$$X - \begin{pmatrix} W_{11} + W_{21}^T W_{22}^{-1} W_{21} & W_{21}^T W_{22}^{-1} \\ W_{22}^{-1} W_{21} & W_{22}^{-1} \end{pmatrix} \succ 0. \tag{32}$$

Therefore, W_{22} and hence also Y are non-singular. The non-singular matrix on the left in (32) is nothing but $X - Y^{-1}Z$. Therefore $Z - YX$ is non-singular. This allows to choose square and non-singular matrices U and V with $VU^T = Z - YX$. We can then define the square and non-singular matrix $\mathcal{Y}$ as in (24) and solve the second equation in (24) for the (necessarily symmetric) matrix $\mathcal{X}$. Similarly we can solve (26) for A_K, B_K, C_K and D_K which specify a controller K. With these definitions all the relations used in the necessity part of the proof hold true, and we can just reverse the arguments. Indeed (27) is identical to (25); since $\mathcal{Y}$ is square and non-singular, a congruence transformation with diag($\mathcal{Y},I$) leads to (18). Moreover, there exists an $\hat{M}$ with $W_{11}+\hat{M} \succ 0$; then (31) implies (30) and thus (29) and hence, by non-singularity of $\mathcal{Y}$, again (20). This completes the proof of Theorem 1.

Note that the last part of the proof provides a constructive procedure for designing a controller. It is easily verified that the McMillan degree of K is not larger than the dimension of A which is the sum of the dimensions of A_H (the degree of H) and of A_Ψ (the degree of Ψ).

We stress that the proposed convexifying transformation can be viewed as a one-shot combination of those in [18, 13] and [16] as already exploited in [8] and [21]. It is worthwhile to emphasize that the transformation of [18, 13] directly applies without complication if Ψ involves no dynamics. Due to our generic problem formulation we fully cover weighted robust estimator [19] and weighted robust feed-forward synthesis involving dynamic IQCs in a unifying manner. In comparison to this previous work, we would like to highlight that we made effective use of the fact that the the (possibly unstable) dynamics in Ψ are outside the control loop which leads to a considerable simplifications of the proofs in this paper under minimal hypotheses.

4.3 Elimination

This little section serves to illustrate how we can routinely eliminate (possibly high-dimensional) variables in our general synthesis result. For this purpose let us note that, due to (16),

$$(0 \ I) \boldsymbol{D}(N) = (0 \ I) \left(D_1 + (E \ 0) N \begin{pmatrix} 0 \\ F \end{pmatrix} \right) = D_{H1} + (D_{H12} \ 0) N \begin{pmatrix} 0 \\ F \end{pmatrix}.$$

If we also recall (28) and $ZB = B$ we conclude that (27) can be written as

$$\boldsymbol{E}(M,W,X) + \mathrm{He} \begin{pmatrix} B & 0 \\ 0 & I \\ \overline{D_{H12} \ 0} \end{pmatrix} N \begin{pmatrix} I & 0 & 0 \\ 0 & C & F \end{pmatrix} \prec 0$$

with some easily determined map $\boldsymbol{E}$ that is affine in (M,W,X). This allows us to apply the projection lemma [9] and eliminate N. In order to formulate the resulting synthesis LMIs we introduce basis matrices $\boldsymbol{F}$ and $\boldsymbol{G}$ of, respectively, the kernels of

$$\begin{pmatrix} B^T & 0 & D_{H12}^T \\ 0 & I & 0 \end{pmatrix} \quad \text{and} \quad \begin{pmatrix} I & 0 & 0 \\ 0 & C & F \end{pmatrix}. \tag{33}$$

This leads to the following synthesis result with reduced computational complexity.

Corollary 1. *There exists* $M \in \mathcal{M}$ *and solutions* (N,W,X) *of the LMIs (27) and (31) if and only if there exist* $M \in \mathcal{M}$ *and* (W,X) *satisfying the LMIs (31) as well as*

$$\boldsymbol{F}^T \boldsymbol{E}(M,W,X)\boldsymbol{F} \prec 0 \quad \text{and} \quad \boldsymbol{G}^T \boldsymbol{E}(M,W,X)\boldsymbol{G} \prec 0.$$

For actual coding, these LMIs can be rendered somewhat more explicit by exploiting the particular structures of B, C, $F = D_{H21}$ and of the matrices (33) in order to determine tailored basis matrices of the kernels of (33).

Technically this leads to the results of [11] (for stable Ψ) and similar ones in [17] that have been obtained independently. Conceptually it fully covers the robust feedforward synthesis results in [12] with a substantially simplified proof and for more general configurations. Note that a similar unifying approach is possible for H_2-synthesis which is not pursued in this paper for reasons of space.

5 A Sketch of Further Applications

Generalized l_2-synthesis has been introduced in [5] in order to handle nominal performance synthesis with independently norm-bounded disturbances or robust synthesis against full but element-by-element bounded uncertainties. With LMIable sets
of real matrices $\mathscr{L}$ and $\mathscr{R}$ this has been translated to the question of finding a controller K which stabilizes H and such that there exist $L \in \mathscr{L}$ and $R \in \mathscr{R}$ with

$$\|L^{-1/2}(H \star K)R^{-1/2}\|_\infty < 1 \quad \text{or} \quad \begin{pmatrix} L & H \star K \\ (H \star K)^* & R \end{pmatrix} \succ 0 \quad \text{on } \mathbb{C}^0.$$

Convexification by controller transformation can be achieved with [18, 13] while
[5] presents the LMI synthesis condition obtained after elimination. An extension
to cost criteria and disturbances described by dynamic IQCs has been discussed in
[11, 6], with applications to e.g. the suppression of disturbances with specific spec-
tral contents. In [6] it is shown that this amounts to the above described scenario for
LMIable sets $\mathscr{L}$ and $\mathscr{R}$ of transfer matrices, and inner-outer factorizations are employed to reduce the synthesis problem to one for static multipliers. More generally,
[11] proposes a more direct solution to synthesizing a controller K which stabilizes
H and for which there exists some $M \in \mathscr{M}$ with

$$\Psi^* M \Psi + \begin{pmatrix} I \\ H \star K \end{pmatrix}^* \Pi_p \begin{pmatrix} I \\ H \star K \end{pmatrix} \prec 0 \quad \text{on } \mathbb{C}^0.$$

This FDI translates into

$$\begin{pmatrix} Q_p + \Psi^* M \Psi & 0 \\ 0 & -I \end{pmatrix} + \mathrm{He}\left(\begin{pmatrix} S_p \\ T_p \end{pmatrix} (H \star K)(I \ 0) \right) \prec 0 \quad \text{on } \mathbb{C}^0,$$

which clarifies that all these design questions subsume to the more specifically looking but as general problem in Section 2.

Techniques that are directly related to the ones presented here have been applied
in [15]. Given stable transfer matrices H_0, H_1, H_2 and a generalized plant H, this
paper covers the design of a controller K which stabilizes H and for which there
exists some $M \in \mathscr{M}$ with

$$\|H_0 + H_1 M H_2 - H \star K\|_\infty < \gamma. \tag{34}$$

It is shown that this encompasses a rich class of questions related to structured controller synthesis and multi-objective control, see also [17]. Mathematically it corresponds to the projection of the transfer matrix of a controlled system onto an affinely
parameterized set of transfer matrices in the H_∞-norm. Since (34) is nothing but

$$\begin{pmatrix} -\gamma I & G_0 + G_1 M G_2 \\ (G_0 + G_1 M G_2)^* & -\gamma I \end{pmatrix} + \mathrm{He}\begin{pmatrix} -I \\ 0 \end{pmatrix} (G \star K)(0 \ I) \prec 0 \quad \text{on } \mathbb{C}^0,$$

this is yet another variant of the problem in Section 2.

6 Conclusions

In this paper we have provided a complete solution to a general parametric dynamic frequency-domain feasibility problem under minimal hypothesis. We have as well worked out various relations to nominal and robust controller synthesis problems in the literature. In particular it has been shown that robust synthesis for generalized plants whose control channel is not affected by uncertainties can be turned into a semi-definite optimization problem.

Acknowledgements The author appreciates many stimulating discussions with Jan Willems (Katholieke Universiteit Leuven), Emre Köse (Boğazici University Istanbul), Hakan Köroğlu and Sjoerd Dietz (Delft University of Technology). Moreover, financial support from MicroNed, a program to promote micro-systems-technology in The Netherlands, is gratefully acknowledged.

7 Appendix: Proof of Lemma 1

Let us partition Π_2^{-1} similarly as Π_2 into the blocks $\tilde{Q}_2$, $\tilde{R}_2$ and $\tilde{S}_2$ and recall that

$$
\begin{pmatrix} \tilde{Q}_2 & \tilde{S}_2 \\ \tilde{S}_2^* & \tilde{R}_2 \end{pmatrix} = \begin{pmatrix} I & 0 \\ -R_2^{-1}S_2 & I \end{pmatrix} \begin{pmatrix} (Q_2 - S_2 R_2^{-1} S_2^*)^{-1} & 0 \\ 0 & R_2^{-1} \end{pmatrix} \begin{pmatrix} I & -S_2^* R_2^{-1} \\ 0 & I \end{pmatrix}.
$$

Let us remember that $R_2 \succcurlyeq 0$ on $\mathbb{C}^0$. We first give the proof under the assumption $R_2 \succ 0$ on $\mathbb{C}^0$. If we abbreviate

$$
T_1 := \begin{pmatrix} I & 0 & 0 \\ P_{11} & 0 & P_{13} \end{pmatrix}^* \Pi_1 \begin{pmatrix} I & 0 & 0 \\ P_{11} & 0 & P_{13} \end{pmatrix}
$$

and recall (5), the inequality (9) can be written as

$$
T_1 + \begin{pmatrix} 0 & I & 0 \\ \mathscr{P}_{21} & P_{22} + R_2^{-1} S_2 & \mathscr{P}_{23} \end{pmatrix}^* \begin{pmatrix} \tilde{Q}_2^{-1} & 0 \\ 0 & R_2 \end{pmatrix} \begin{pmatrix} 0 & I & 0 \\ \mathscr{P}_{21} & P_{22} + R_2^{-1} S_2 & \mathscr{P}_{23} \end{pmatrix} +
$$
$$
+ \begin{pmatrix} 0 & 0 & \mathscr{P}_{31}^* S_p^T \\ 0 & 0 & P_{32}^* S_p^T \\ S_p \mathscr{P}_{31} & S_p P_{32} & S_p \mathscr{P}_{33} + Q_p + \mathscr{P}_{33}^* S_p^T \end{pmatrix} +
$$
$$
+ \begin{pmatrix} \mathscr{P}_{31} & P_{32} & \mathscr{P}_{33} \end{pmatrix}^* T_p^T T_p \begin{pmatrix} \mathscr{P}_{31} & P_{32} & \mathscr{P}_{33} \end{pmatrix} \prec 0
$$

and thus (Schur)

$$\left(T_1 + \begin{pmatrix} \begin{pmatrix} 0 & 0 & 0 \\ 0 & \tilde{Q}_2^{-1} & P_{32}^* S_p^T \\ 0 & S_p P_{32} & Q_p \end{pmatrix} & \begin{pmatrix} 0 \\ P_{22}^* - \tilde{Q}_2^{-1}\tilde{S}_2 \\ 0 \end{pmatrix} & \begin{pmatrix} 0 \\ P_{32}^* T_p^T \\ 0 \end{pmatrix} \\ \begin{pmatrix} 0 & P_{22} - \tilde{S}_2^* \tilde{Q}_2^{-1} & 0 \end{pmatrix} & -R_2^{-1} & 0 \\ \begin{pmatrix} 0 & T_p P_{32} & 0 \end{pmatrix} & 0 & -I \end{pmatrix} + T_2 \prec 0 \quad (35)$$

with

$$T_2 := \mathrm{He} \begin{pmatrix} 0 & 0 \\ 0 & 0 \\ 0 & S_p \\ I & 0 \\ 0 & T_p \end{pmatrix} \begin{pmatrix} \mathscr{P}_{21} & \mathscr{P}_{23} \\ \mathscr{P}_{31} & \mathscr{P}_{33} \end{pmatrix} \begin{pmatrix} I & 0 & 0 & 0 & 0 \\ 0 & 0 & I & 0 & 0 \end{pmatrix}.$$

Note that the second block rows and columns of T_1 and T_2 vanish; therefore we can eliminate all non-zero off-diagonal blocks in the second row/column of (35) without affecting T_1 and T_2 at all; this elimination amounts to performing a congruence transformation with a transfer matrix that is upper block-triangular and has identity blocks on its diagonal. For

$$\begin{pmatrix} \begin{pmatrix} 0 & 0 & 0 \\ 0 & \tilde{Q}_2^{-1} & P_{32}^* S_p^T \\ 0 & S_p P_{32} & Q_p \end{pmatrix} & \begin{pmatrix} 0 \\ P_{22}^* - \tilde{Q}_2^{-1}\tilde{S}_2 \\ 0 \end{pmatrix} & \begin{pmatrix} 0 \\ P_{32}^* T_p^T \\ 0 \end{pmatrix} \\ \begin{pmatrix} 0 & P_{22} - \tilde{S}_2^* \tilde{Q}_2^{-1} & 0 \end{pmatrix} & -R_2^{-1} & 0 \\ \begin{pmatrix} 0 & T_p P_{32} & 0 \end{pmatrix} & 0 & -I \end{pmatrix}$$

this operation is easily seen to result in

$$\begin{pmatrix} 0 & 0 & 0 & 0 & 0 \\ 0 & \tilde{Q}_2^{-1} & 0 & 0 & 0 \\ 0 & 0 & Q_p & 0 & 0 \\ 0 & 0 & 0 & 0 & 0 \\ 0 & 0 & 0 & 0 & -I \end{pmatrix} - \begin{pmatrix} 0 & 0 \\ 0 & 0 \\ 0 & S_p \\ I & 0 \\ 0 & T_p \end{pmatrix} \begin{pmatrix} P_{22} & -I \\ P_{32} & 0 \end{pmatrix} \tilde{\Pi}_2 \begin{pmatrix} P_{22} & -I \\ P_{32} & 0 \end{pmatrix}^* \begin{pmatrix} 0 & 0 \\ 0 & 0 \\ 0 & S_p \\ I & 0 \\ 0 & T_p \end{pmatrix}^T.$$

After canceling the second row and column, we have proved that (9) implies (11).

Now consider the case that R_2 is not positive definite on $\mathbb{C}^0$. Then (9) persists to hold if replacing the right-hand side of (9) by $-\delta I$ for some sufficiently small $\delta > 0$. We can then repeat all arguments after perturbing R_2 to $R_2 + \varepsilon I$ for any $\varepsilon > 0$. Let us denote the corresponding multiplier by Π_2^ε and apply the above described procedure. It is crucial to observe that $(\Pi_2^\varepsilon)^{-1}$ converges to Π_2^{-1} for $\varepsilon \to 0$ since Π_2 is non-singular. Denote the upper triangular elimination matrix appearing in the course of the arguments by T_ε; observe that T_ε has ones on its diagonal and that its upper-diagonal elements which depend on ε only involve blocks from $(\Pi_2^\varepsilon)^{-1}$; hence T_ε is guaranteed to converge, for $\varepsilon \to 0$, to a matrix T_0 which is non-singular on $\mathbb{C}^0$. The given proof then leads to (11) with Π_2^{-1} replaced by $(\Pi_2^\varepsilon)^{-1}$ and the right-hand side being $-\delta T_\varepsilon^* T_\varepsilon$. This allows to take the limit $\varepsilon \to 0$ and we obtain (11) again.

The converse statement is proved by first assuming $\tilde{Q}_2 \prec 0$ and then using a perturbation argument similar to the one just provided.

References

[1] Balakrishan, V., Vandenberghe, L.: Semidefinite programming duality and linear time-invariant systems. IEEE Trans. Aut. Contr. **48**(1), 30–41 (2003)

[2] Ben-Tal, A., Nemirovski, A.: Lectures on Modern Convex Optimization: Analysis, Algorithms, and Engineering Applications. SIAM-MPS Series in Optimizaton. SIAM Publications, Philadelphia (2001)

[3] Boyd, S., El Ghaoui, L., Feron, E., Balakrishan, V.: Linear matrix inequalities in system and control theory. SIAM Studies in Applied Mathematics 15. SIAM, Philadelphia (1994)

[4] Cerone, V., Milanese, M., Regruto, D.: Robust feedforward design for a two-degrees of freedom controller. Systems & Control Letters **56**(11-12), 736–741 (2007)

[5] D'Andrea, R.: Generalized l_2 synthesis. IEEE Trans. Aut. Contr. **44**(6), 1145–1156 (1999)

[6] D'Andrea, R.: Convex and finite-dimensional conditions for controller synthesis with dynamic integral constraints. IEEE Transactions on Automatic Control **46**(2), 222–234 (2001)

[7] Desoer, C., Vidyasagar, M.: Feedback Systems: Input-Output Approach. Academic Press, London (1975)

[8] Dietz, S.: Analysis and control of uncertain systems by using robust semi-definite programming. Ph.D. thesis, Delft University of Technology (2008)

[9] Gahinet, P., Apkarian, P.: A linear matrix inequality approach to H_∞ control. Int. J. Robust Nonlin. **4**, 421–448 (1994)

[10] Giousto, A., Paganini, F.: Robust synthesis of feedforward compensators. IEEE Trans. Aut. Contr. **44**(8), 1578–1582 (1999)

[11] Kao, C.Y., Ravuri, M., Megretski, A.: Control synthesis with dynamic integral quadratic constraints - LMI approach. In: Proc. 39th IEEE Conf. Decision and Control, pp. 1477–1482. Sydney, Australia (2000)

[12] Köse, I.E., Scherer, C.W.: Robust L_2-gain feedforward control of uncertain systems using dynamic IQCs. International Journal of Robust and Nonlinear Control p. 24 (2009)

[13] Masubuchi, I., Ohara, A., Suda, N.: LMI-based controller synthesis: a unified formulation and solution. Int. J. Robust Nonlin. **8**, 669–686 (1998)

[14] Megretski, A., Rantzer, A.: System analysis via integral quadratic constraints. IEEE T. Automat. Contr. **42**, 819–830 (1997)

[15] Scherer, C.W.: Design of structured controllers with applications. In: Proc. 39th IEEE Conf. Decision and Control. Sydney, Australia (2000)

[16] Scherer, C.W.: An efficient solution to multi-objective control problems with LMI objectives. Syst. Contr. Letters **40**(1), 43–57 (2000)

[17] Scherer, C.W.: Multi-objective control without Youla parameterization. In: S. Moheimani (ed.) Perspectives in robust control, *Lecture Notes in Control and Information Sciences*, vol. 256, pp. 311–325. Springer-Verlag, London (2001)

[18] Scherer, C., Gahinet, P., Chilali, M.: Multi-objective output-feedback control via LMI optimization. IEEE T. Automat. Contr. **42**, 896–911 (1997)

[19] Scherer, C.W., Köse, I.E.: Robustness with dynamic IQCs: An exact state-space characterization of nominal stability with applications to robust estimation. Automatica **44**(7), 1666–1675 (2008)

[20] Sun, K.P., Packard, A.: Robust H_2 and H_∞ filters for uncertain LFT systems. IEEE Trans. Aut. Contr. **50**(5), 715–720 (2005)

[21] Veenman, J., Köroğlu, H., Scherer, C.W.: An IQC approach to robust estimation against perturbations of smoothly time-varying parameters. In: Proc. 47th IEEE Conf. Decision and Control. Cancun, Mexico (2008)

Conservation Laws and Lumped System Dynamics

Arjan van der Schaft, Bernhard Maschke

1 Introduction

Physical systems modeling, aimed at network modeling of complex multi-physics systems, has especially flourished in the fifties and sixties of the 20-th century, see e.g. [11, 12] and references provided therein. With the reinforcement of the 'systems' legacy in Systems & Control, the growing recognition that 'control' is not confined to developing algorithms for processing the measurements of the system into control signals (but instead is concerned with the design of the total controlled system), and facing the complexity of modern technological and natural systems, systematic methods for physical systems modeling of large-scale lumped- and distributed-parameter systems capturing their basic physical characteristics are needed more than ever.

In this paper we are concerned with the development of a systematic framework for modeling multi-physics systems which is directly based on *conservation laws*. Modeling based on conservation laws is prevalent in a distributed-parameter context in areas such as fluid dynamics and hydraulic systems, chemical and thermodynamical systems [2], as well as electromagnetism, but is also underlying the basic structure of lumped-parameter systems such as electrical circuits. While the natural framework for formulating Kirchhoff's laws for electrical circuits is the circuit *graph* we will show in this paper how distributed-parameter conservation laws can be discretized by using the proper generalization of the notion of graph to 'higher-dimensional networks', called *k-complexes* in algebraic topology. Furthermore, we show how these discretized conservation laws define a power-conserving intercon-

---

A.J. van der Schaft
Institute of Mathematics and Computing Science, University of Groningen, PO Box 407, 9700 AK, the Netherlands, e-mail: A.J.van.der.Schaft@rug.nl

B.M. Maschke
Lab. d'Automatique et de Genie des Procédés, Université Claude Bernard Lyon-1, F-69622 Villeurbanne, Cedex, France, e-mail: maschke@lagep.univ-lyon1.fr

P.M.J. Van den Hof et al. (eds.), *Model-Based Control: Bridging Rigorous Theory and Advanced Technology*, DOI: 10.1007/978-1-4419-0895-7_3,
© Springer Science + Business Media, LLC 2009

nection structure, called a Dirac structure, which, when combined with the (discretized) constitutive relations, defines a finite-dimensional *port-Hamiltonian system* [14, 13, 5].

In previous work [15] we have laid down a framework for formulating conservation laws described by partial differential equations as infinite-dimensional port-Hamiltonian systems. Furthermore, in [8] we have shown how such infinite-dimensional port-Hamiltonian systems can be spatially discretized to finite-dimensional port-Hamiltonian systems by making use of mixed finite-element methods. In this paper we show how alternatively we can directly spatially 'lump' the dynamics described by conservation laws in a structure-preserving manner, again obtaining a finite-dimensional port-Hamiltonian system description. This approach also elucidates the concept of the spatial system boundary, and leads to the notion of *distributed terminals*.

This paper is a follow-up of our previous paper [16]. Older references in this spirit include [10, 12].

2 Kirchhoff's laws on graphs and circuit dynamics

In this section we recall the abstract formulation of Kirchhoff's laws on graphs, dating back to the historical work of Kirchhoff [9], as can be found e.g. in [1, 3]. In order to deal with *open* electrical circuits we define *open graphs*, and we show how Kirchhoff's laws on open graphs define a power-conserving interconnection structure, called a Dirac structure, between the currents through and the voltages over the edges of the graph, and the boundary currents and potentials. This enables us to describe the circuit dynamics as a port-Hamiltonian system.

2.1 Graphs

An *oriented graph*[1] $\mathcal{G}$, see e.g. [3], consists of a finite set $\mathcal{V}$ of *vertices* and a finite set $\mathcal{E}$ of directed *edges*, together with a mapping from $\mathcal{E}$ to the set of ordered pairs of $\mathcal{V}$. Thus to any branch $e \in \mathcal{E}$ there corresponds an ordered pair $(v,w) \in \mathcal{V}^2$ representing the initial vertex v and the final vertex w of this edge. An oriented graph is completely specified by its *incidence matrix B*, which is an $\bar{v} \times \bar{e}$ matrix, $\bar{v}$ being the number of vertices and $\bar{e}$ being the number of edges, with (i,j)-th element b_{ij} equal to 1 if the j-th edge is an edge towards vertex i, equal to -1 if the j-th edge is an edge originating from vertex i, and 0 otherwise.

Given an oriented graph we define its *vertex space* Λ_0 as the real vector space of all functions from $\mathcal{V}$ to $\mathbb{R}$. Clearly Λ_0 can be identified with $\mathbb{R}^{\bar{v}}$. Furthermore, we

---

[1] In fact, we will be considering *multi-graphs* since we allow for the existence of multiple branches between the same pair of vertices.

define its *edge space* Λ_1 as the vector space of all functions from $\mathscr{E}$ to $\mathbb{R}$. Again, Λ_1 can be identified with $\mathbb{R}^{\tilde{e}}$.

In the context of an electrical circuit Λ_1 will be the vector space of currents *through* the edges in the circuit. The dual space of Λ_1 will be denoted by Λ^1, and defines the vector space of voltages *across* the edges. (We have highlighted the words 'through' and 'across' to refer to the classical use of 'through' and 'across' variables, see e.g. [11].) Furthermore, the duality product $< V|I > = V^T I$ of a vector of currents $I \in \Lambda_1$ with a vector of voltages $V \in \Lambda^1$ is the total power over the circuit. Similarly, the dual space of Λ_0 is denoted by Λ^0 and defines the vector space of potentials at the vertices.

Remark 1. Since Λ_0 and Λ_1 have a canonical basis corresponding to the individual vertices, respectively edges, there is a standard Euclidean inner product on both spaces, and thus both Λ^0 and Λ^1 can be identified with Λ_0, respectively Λ_1, such that the duality product becomes this standard inner product. In situations to be treated later on this will not necessarily be the case.

The incidence matrix B can be also regarded as the matrix representation of a linear map (denoted by the same symbol)

$$B : \Lambda_1 \rightarrow \Lambda_0$$

called the *incidence operator* or (*boundary operator*). Its adjoint map is denoted in matrix representation as

$$B^T : \Lambda^0 \rightarrow \Lambda^1,$$

and is called the *co-incidence* (or *co-boundary*) operator.

2.2 Kirchhoff's laws for graphs

Consider an oriented graph $\mathscr{G}$ specified by its incidence operator B. Kirchhoff's laws associated with the graph are expressed as follows. Kirchhoff's current laws (KCL) are given as

$$I \in \mathbf{kernel} B, \tag{1}$$

while Kirchhoff's voltage laws (KVL) take the form

$$V \in \mathrm{im} B^T. \tag{2}$$

A graph theoretic interpretation of Kirchhoff's current and voltage laws can be given as follows [3]. The kernel of the incidence operator B is the *cycle space* $Z \subset \Lambda_1$ of the graph, while the image $U \subset \Lambda^1$ of the co-incidence operator B^T is its *cut space* (or, co-cycle space). Since $\mathbf{kernel} B = (\mathrm{im} B^T)^\perp$ (with $\perp$ denoting the orthogonal complement with respect to the duality product between the dual spaces Λ_1 and Λ^1) the cycle space is the orthogonal complement of the cut space.

This leads to the equivalent way of formulating Kirchhoff's current laws as the fact that the total current I along any cut is equal to zero, since $I \in \mathbf{kernel}B$ is equivalent to I being orthogonal to the cut space U. The simplest elements of the cut space U (which in fact are spanning the linear space U) are the cuts given by all edges starting from or terminating on a single vertex v. Kirchhoff's current laws for these cut sets mean nothing else than the expression that the currents entering or leaving any vertex v sum up to zero. Indeed, if v is numbered as the i-th vertex then the i-the equation in the linear set of equations $BI = 0$ is precisely this.

On the other hand, since $V \in \mathrm{im}\,B^T$ is equivalent to V being orthogonal to the cycle space Z, Kirchhoff's voltage laws can be equivalently described as the fact that the total voltage over every cycle is zero.

The difference between Kirchhoff's current and voltage laws is also reflected by writing Kirchhoff's voltage laws as

$$V = B^T \psi \tag{3}$$

for some vector $\psi \in \Lambda^0$, which has the physical interpretation of being the vector of *potentials* at every vertex. Hence Kirchhoff's voltage laws express that the voltage distribution V over the edges of the graph corresponds to a potential distribution over the vertices.

Of course, Tellegen's theorem automatically follows from Kirchhoff's laws. Indeed, take any current distribution I satisfying Kirchhoff's current laws $BI = 0$, and any voltage distribution V satisfying Kirchhoff's voltage laws $V = B^T \psi$. Then,

$$V^T I = \psi^T BI = 0 \tag{4}$$

In particular, Tellegen's theorem implies that for any *actual* current and voltage distribution over the circuit the total power $V^T I$ is equal to zero.

We summarize the Kirchhoff behavior $\mathcal{B}_K(\mathcal{G})$ of a graph $\mathcal{G}$ with incidence matrix B as

$$\mathcal{B}_K(\mathcal{G}) := \{(I,V) \in \Lambda_1 \times \Lambda^1 \mid I \in \mathbf{kernel}B, V \in \mathrm{im}\,B^T\} \tag{5}$$

It immediately follows that the Kirchhoff behavior defines a *Dirac structure*. Recall [6, 14, 13] that a subspace $D \subset V \times V^*$ for some vector space V defines a Dirac structure if $D = D^{\mathrm{orth}}$ where ^{orth} denotes the orthogonal complement with respect to the indefinite inner product $<<,>>$ on $V \times V^*$ defined as

$$<< (v_1, v_1^*), (v_2, v_2^*) >>:=< v_1^* | v_2 > + < v_2^* | v_1 >$$

with $v_1, v_2 \in V, v_1^*, v_2^* \in V^*$, where $<|>$ denotes the duality product between V and V^*.

2.3 Kirchhoff's laws for open graphs

Although in Kirchhoff's original treatment of circuits and graphs external currents entering the vertices of the graph were an indispensable notion, this has not been articulated very well in the subsequent formalization of circuits and graphs[2]. Hence a reinforcement of this *systems point of view* is definitely in order.

We will do so by extending the notion of graph to *open graph*. An open graph $\mathscr{G}$ is obtained from an ordinary graph with set of vertices $\mathcal{V}$ by identifying a subset $\mathcal{V}_b \subset \mathcal{V}$ of *boundary vertices*. The interpretation of $\mathcal{V}_b$ is that these are the vertices that are open to interconnection (e.g., with other graphs). The remaining subset $\mathcal{V}_i :=$ $\mathcal{V} - \mathcal{V}_b$ are the *internal vertices* of the open graph.

Remark 2. Another way of defining open graphs is by identifying some of the *edges* to be the *boundary edges* (open to interconnection). Such a definition is straightforward, and we will not elaborate on this. The distinction between the definitions of an open graph using boundary vertices or boundary edges is analogous to the difference between *boundary control* of distributed-parameter systems and *distributed control*; see also Section 3.

Kirchhoff's current laws apply to an open graph $\mathscr{G}$ in a different manner than to an ordinary graph, since the ordinary Kirchhoff's current laws would imply that the sum of the currents over all edges incident on a boundary vertex is *zero*, which is *not* what we want for interconnection. Furthermore, by Tellegen's theorem, the ordinary KCL would imply that the total power in the circuit is equal to zero, thus implying that there cannot be any ingoing or outgoing power flow. Hence we have to modify Kirchhoff's current laws by requiring that the incidence operator B maps the vector of currents I to a vector that has zero components corresponding to the internal vertices, while for the boundary vertices the image is equal to (minus) the vector of *boundary currents* I_b. Decomposing the incidence operator B as $\begin{bmatrix} B_i \\ B_b \end{bmatrix}$ with B_i the part of the incidence operator corresponding to the internal vertices, and B_b the part corresponding to the boundary vertices, we thus arrive at

$$B_i I = 0, \quad B_b I = -I_b, \quad \text{KCL} \tag{6}$$

Here the vector I_b is belonging to the vector space Λ_b of functions from the boundary vertices $\mathcal{V}_b$ to $\mathbb{R}$ (which is identified with $\mathbb{R}^{\bar{v}_b}$, with $\bar{v}_b$ the number of boundary vertices)[3].

Kirchhoff's voltage laws (KVL) remain unchanged, and will be written as

---

[2] Unfortunately, this holds for many formalizations of physical theories over the last century. A proper theory of *mechanics* should include external forces from the very start, instead of restricting itself to closed mechanical systems. *Thermodynamics* cannot be properly formalized without taking interaction with other systems into account.

[3] Alternatively, open graphs can be defined by attaching 'one-sided open edges' (properly called *leaves*) to every boundary vertex in $\mathcal{V}_b$. Then the elements of the vector I_b are the currents through these leaves, see also [17].

$$V = B^T \psi = B_i^T \psi_i + B_b^T \psi_b, \quad \text{KVL} \tag{7}$$

where ψ_i denotes the vector of the potentials at the internal vertices and ψ_b the vector of potentials at the boundary vertices. Note that $\psi_b \in \Lambda^b$ (where we define Λ^b to be the dual of the space of boundary currents Λ_b). This results in the following Kirchhoff behavior for an open graph $\mathscr{G}$:

$$B_K(\mathscr{G}) := \{(I, V, I_b, \psi_b) \in \Lambda_1 \times \Lambda^1 \times \Lambda_b \times \Lambda^b \mid \tag{8}$$

$$B_i I = 0, B_b I = -I_b, \exists \psi_i \text{ s.t. } V = B_i^T \psi_i + B_b^T \psi_b\}$$

By computing as before, cf. (4), the total power over the graph we now obtain

$$V^T I = \psi_i^T B_i I + \psi_b^T B_b I = -\psi_b^T I_b \tag{9}$$

Thus, for open graphs the total power $V^T I$ is equal to the outgoing power $-\psi_b^T I_b$. This will lead to the following characterization of the Kirchhoff behavior of open graphs as Dirac structures.

Proposition 1. *Let $\mathscr{G}$ be an open graph with incidence matrix $B = \begin{bmatrix} B_i \\ B_b \end{bmatrix}$. Then its Kirchhoff behavior $\mathscr{B}_K(\mathscr{G})$ is a Dirac structure.*

Proof. As shown in (9), $V^T I + \psi_b^T I_b = 0$. By considering $I = I_1 + I_2, V = V_1 + V_2, I_b = I_{b1} + I_{b2}, \psi = \psi_1 + \psi_2$, with $(I_j, V_j, I_{bj}, \psi_j) \in \mathscr{B}_K(\mathscr{G}), j = 1, 2$, it follows that

$$\ll (I_1, V_1, I_{b1}, \psi_1), (I_2, V_2, I_{b2}, \psi_2) \gg = 0,$$

which implies that $\mathscr{B}_K(\mathscr{G}) \subset (\mathscr{B}_K(\mathscr{G}))^{\text{orth}}$.
For showing the reverse inclusion, consider a quadruple $(I, V, I_b, \psi_b) \in (\mathscr{B}_K(\mathscr{G}))^{\text{orth}}$, that is,

$$V^T \bar{I} + \bar{V}^T I + \psi_i^T \bar{I}_b + \bar{\psi}_b^T I_b = 0 \tag{10}$$

for all $(\bar{I}, \bar{V}, \bar{I}_b, \bar{\psi}_b)$ satisfying

$$B_i \bar{I} = 0, B_b \bar{I} = -\bar{I}_b, \bar{V} = B_i^T \bar{\psi}_i + B_b^T \bar{\psi}_b, \quad \text{for some } \bar{\psi}_i.$$

Writing out (10) we obtain

$$0 = V^T \bar{I} + \bar{\psi}_i^T B_i I + \bar{\psi}_b^T B_b I - \bar{\psi}_b^T B_i I + \bar{\psi}_b^T I_b$$

$$= (V - B_b^T \psi_b)^T \bar{I} + \bar{\psi}_i^T B_i I + \bar{\psi}_b^T (B_b I + I_b)$$

for all $\bar{\psi}_i, \bar{\psi}_b$ and all $\bar{I}$ satisfying $B_i \bar{I} = 0$. It follows that $B_i I = 0$, $B_b I + I_b = 0$, and $V - B_b^T \psi_b \in \text{im} B_i^T$, or equivalently $V - B_b^T \psi_b = B_i^T \psi_i$ for some ψ_i. $\qquad\square$

2.4 Constraints on boundary currents and invariance of boundary potentials

It is a well-known property [3] of any incidence matrix B that

$$\mathbb{1}^T B = 0 \tag{11}$$

where $\mathbb{1}$ denotes the vector with all components equal to 1. From this property it follows that the rank of the incidence matrix B is at most $\bar{v} - 1$. In fact, the rank is given as [3] $\operatorname{rank} B = \bar{v} - c_{\mathscr{G}}$, where $c_{\mathscr{G}}$ is the number of components[4] of the graph $\mathscr{G}$. (Thus $\operatorname{rank} B = \bar{v} - 1$ for a connected graph.) By (6) it follows that

$$0 = \mathbb{1}^T BI = \mathbb{1}_b^T B_b I = -\mathbb{1}_b^T I_b = -\sum_{v_b} I_{v_b} \tag{12}$$

with $\mathbb{1}_b$ denoting the vector with all ones of dimension equal to the number of boundary vertices, and where the summation is over all boundary vertices $v_b \in \mathscr{V}_b$. Hence the boundary part of the Kirchhoff behavior of an open graph is constrained by the obvious fact that all boundary currents sum up to zero. Dually, we may always add to the vector of potentials ψ the vector $\mathbb{1}$ leaving invariant the vector of voltages $V = B^T \psi$. Hence, to the vector of boundary potentials ψ_b we may always add the vector $\mathbb{1}_b$. Summarizing we arrive at a similar statement as in [17]):

Proposition 2. *Consider an open graph $\mathscr{G}$ with Kirchhoff behavior $\mathscr{B}_K(\mathscr{G})$. Then for each $(I, V, I_b, \psi_b) \in \mathscr{B}_K(\mathscr{G})$ it holds that*

$$\mathbb{1}_b^T I_b = 0$$

while for any constant $c \in \mathbb{R}$

$$(I, V, I_b, \psi_b + c\mathbb{1}_b) \in \mathscr{B}_K(\mathscr{G})$$

This implies that we may restrict the dimension of the space of external variables $\Lambda_b \times \Lambda^b$ by two. Indeed, we may define

$$\Lambda_{bred} := \{I_b \in \Lambda_b \mid I_b \in \mathbf{kernel}\, \mathbb{1}_b^T\}$$

and its dual space

$$\Lambda_{red}^b := (\Lambda_{bred})^* = \Lambda^b / \operatorname{im} \mathbb{1}_b$$

It is rather straightforward to show that the Kirchhoff behavior $\mathscr{B}_K(\mathscr{G})$ reduces to a linear subspace of the reduced space $\Lambda_1 \times \Lambda^1 \times \Lambda_{bred} \times \Lambda_{red}^b$, which is also a Dirac structure. A circuit interpretation of this reduction is that we may consider one of the boundary vertices, say the first one, to be the reference ground vertex, and that we may reduce the vector of boundary potentials $\psi_b = (\psi_{b1}, \cdots, \psi_{b\bar{v}_b})$ to a vector

---

[4] A component is a maximal subgraph which is connected, that is, every two vertices are linked by a path of, -non-oriented-, edges.

of voltages $(\psi_{b2} - \psi_{b1}, \cdots, \psi_{b\bar{v}_b} - \psi_{b1})$. A graph-theoretical interpretation is that instead of the incidence operator B we consider the *restricted* incidence operator [1].

For a graph $\mathscr{G}$ with more than one connected component the above holds for each connected component. It follows that there are as many independent constraints on the boundary currents I_b as the number of the connected components of the open graph $\mathscr{G}$. Dually, the space of allowed boundary potentials ψ_b is invariant under translation by as many independent vectors $\mathbb{1}_b$ as the number of connnected components.

A complementary view on the fact that the sum of the boundary currents is equal to zero and the boundary potentials are invariant under translation along $\mathbb{1}_b$ is the fact that we may *close* an open graph $\mathscr{G}$ to an ordinary graph $\bar{\mathscr{G}}$. Consider first the case that $\mathscr{G}$ is connected. Then we may add one virtual ('ground') vertex v_0, and virtual edges from this virtual vertex to every boundary vertex $v_b \in \mathcal{V}_e$, in such a manner that the Kirchhoff behavior of this graph $\bar{\mathscr{G}}$ *extends* the Kirchhoff behavior of the open graph $\mathscr{G}$. In fact, to the virtual vertex v_0 we may associate an arbitrary potential ψ_{v_0} (a ground-potential), and we may rewrite the righthand-side of (9) as (since $\sum_{v_b} I_{v_b} = 0$)

$$-\sum_{v_b}(\psi_{v_b} - \psi_{v_0})I_{v_b} = -\sum_{v_b} V_{v_b}I_{v_b} \tag{13}$$

where $V_{v_b} := \psi_{v_b} - \psi_{v_0}$ and I_{v_b} denotes the voltage across and the current through the virtual edge towards the boundary vertex v_b.

If the open graph $\mathscr{G}$ consists of more than one component, then one extends the graph by adding a virtual vertex to *every* component containing boundary vertices.

2.5 Interconnection of open graphs

Consider two open graphs $\mathscr{G}^j$ with boundary operators $B^j = \begin{bmatrix} B_i^j \\ B_b^j \end{bmatrix}, j = 1, 2$. *Inter-connection* is done by identifying some of their boundary vertices, and equating (up to a minus sign) the boundary potentials and currents corresponding to these boundary vertices.

For simplicity we consider the case that we can equate *all* their boundary vertices with each other, resulting in an ordinary (closed) graph with set of vertices $\mathcal{V}_i^1 \cup \mathcal{V}_i^2 \cup \mathcal{V}$, where $\mathcal{V}_i := \mathcal{V}_b^1 = \mathcal{V}_b^2$ denotes the set of shared boundary vertices. The incidence operator B of this interconnected graph is given as

$$B = \begin{bmatrix} B_i^1 & 0 \\ 0 & B_i^2 \\ B_b^1 & B_b^2 \end{bmatrix}, \tag{14}$$

corresponding to the following interconnection constraints on the boundary potentials and currents

$$\psi_b^1 = \psi_b^2, \quad I_b^1 + I_b^2 = 0 \tag{15}$$

Of course, several extensions are possible. For example, one may still retain the shared vertices $\mathcal{V}_b := \mathcal{V}_b^1 = \mathcal{V}_b^2$ as being boundary vertices (instead of internal vertices as above) by extending (15) to

$$\psi_b^1 = \psi_b^2 = \psi_b, \quad I_b^1 + I_b^2 + I_b = 0 \tag{16}$$

with I_b, ψ_b the boundary currents and potentials of the interconnected graph.

2.6 Constitutive relations and port-Hamiltonian circuit dynamics

The dynamics of an RLC-circuit is defined, on top of Kirchhoff's laws for its circuit graph, by the constitutive relations of its elements (in this case, capacitors, inductors and resistors). They specify for each edge e a relation between the current I_e through and the voltage V_e across the edge. The simplest case is a *resistive* relation between I_e and V_e such that $V_e I_e \leq 0$. In particular, a linear resistor at edge e is specified by a relation $V_e = -R_e I_e$ with $R_e \geq 0$. In the case of a *capacitive* relation one defines an additional energy variable Q_e (denoting the charge) together with a real function $H_{Ce}(Q_e)$ denoting the electric energy stored in the capacitor. The constitutive relations for a capacitor at edge e are given by

$$\dot{Q}_e = -I_e, \quad V_e = \frac{dH_{Ce}}{dQ_e}(Q_e) \tag{17}$$

Alternatively, in the case of an inductor one specifies the magnetic energy $H_{Le}(\Phi_e)$, where Φ_e denotes the magnetic flux linkage, together with the dynamic relations

$$\dot{\Phi}_e = -V_e, \quad I_e = \frac{dH_{Le}}{d\Phi_e}(\Phi_e) \tag{18}$$

Substituting these constitutive relations into the Kirchhoff behavior $\mathcal{B}_K(\mathcal{G})$ (which is a Dirac structure) results in a port-Hamiltonian[5] system, see e.g. [14, 13], given by

$$((I_C, I_L, I_R), (V_C, V_L, V_R), I_b, \psi_b) \in \mathcal{B}_K(\mathcal{G})$$

where the vectors $I_C, I_L, I_R, V_C, V_L, V_R$ denote the currents, respectively voltages, corresponding to the capacitors, inductors, and resistors, related as

---

[5] Strictly speaking, the terminology 'port-Hamiltonian' is not completely appropriate since this assumes that the system boundary consists of *ports*, that is pairs of vertices with boundary variables being the currents through and the voltages across the edge corresponding to each port. Nevertheless, the mathematical structure and system description remains the same. Furthermore, by reducing the Kirchhoff behavior as above, or alternatively by extending it through the addition of a ground vertex, the boundary variables become true port variables.

$$I_C = -\dot{Q}, V_C = \frac{\partial H_C}{\partial Q}(Q), \quad V_L = -\dot{\Phi}, I_L = \frac{\partial H_L}{\partial \Phi}(\Phi),$$

where Q denotes the vector of charges at the capacitors, Φ denotes the vector of fluxes at the inductors, H_C and H_L denote the total electric and magnetic energies, and where moreover the vectors I_R, V_R satisfy a resistive relation.

Example 1. Let us consider an LC-circuit (for simplicity without boundary vertices). We will start by decomposing the circuit graph $\mathscr{G}$ as the interconnection of a graph corresponding to the capacitors and a graph corresponding to the inductors. Define $\hat{\mathscr{V}}$ as the set of all vertices that are adjacent to at least one capacitor *as well as* to at least one inductor. Then split the circuit graph into an open circuit graph $\mathscr{G}^C$ corresponding to the capacitors and an open circuit graph $\mathscr{G}^L$ corresponding to the inductors, both with set of boundary vertices $\hat{\mathscr{V}}$. Denote the incidence matrices of these two circuit graphs by

$$B^C := \begin{bmatrix} B_i^C \\ B_b^C \end{bmatrix}, B^L := \begin{bmatrix} B_i^L \\ B_b^L \end{bmatrix}$$

Assuming for simplicity that all capacitors and inductors are linear we arrive at the following equations for the C-circuit

$$B_b^C \dot{Q} = I_b^C, B_i^C \dot{Q} = 0$$
$$B_b^{CT} \psi_b^C = C^{-1}Q - B_i^{CT} \psi_i^C$$

with C the diagonal matrix with diagonal elements corresponding to the capacitances of the capacitors, and for the L-circuit

$$\Phi = B_b^{LT} \psi_b^L + B_i^{LT} \psi_i^L$$
$$0 = B_i^L L^{-1} \Phi$$
$$I_b^L = -B_i^L \Phi$$

with L the diagonal matrix of inductances.

The equations of the LC-circuit are obtained by imposing the interconnection constraints $\psi_b^C = \psi_b^L =: \psi_i$ and $I_b^C + I_b^L = 0$. By eliminating the boundary currents I_b^C, I_b^L one arrives at the differential-algebraic equations

$$\begin{bmatrix} B_i^C & 0 \\ 0 & B_i^L \\ B_b^C & B_b^L \end{bmatrix} \begin{bmatrix} -\dot{Q} \\ L^{-1}\Phi \end{bmatrix} = 0$$

$$\begin{bmatrix} C^{-1}Q \\ -\dot{\Phi} \end{bmatrix} = \begin{bmatrix} B_i^{CT} & 0 & B_b^{CT} \\ 0 & B_i^{LT} & B_b^{LT} \end{bmatrix} \begin{bmatrix} \psi_i^C \\ \psi_i^L \\ \psi_i \end{bmatrix}$$

3 Conservation laws on higher-dimensional complexes

In this section we will extend the formalization of conservation laws on graphs as expressed by Kirchhoff's laws to higher-dimensional networks. In particular, this will allow us to systematically spatially *discretize* distributed-parameter physical systems to finite-dimensional lumped-parameter systems, represented in port-Hamiltonian form.

3.1 *Kirchhoff behavior on k-complexes*

An oriented graph with incidence matrix B is a typical example of what is called in algebraic topology a 1-*complex*. Indeed, the sequence

$$\Lambda_1 \xrightarrow{B} \Lambda_0 \xrightarrow{\mathbb{1}} \mathbb{R}$$

satisfies the property $\mathbb{1} \circ B = 0$. In general, a k-complex Λ is specified by a sequence of real linear spaces[6] $\Lambda_0, \Lambda_1, \cdots, \Lambda_k$, together with a sequence of incidence operators

$$\Lambda_k \xrightarrow{\partial_k} \Lambda_{k-1} \xrightarrow{\partial_{k-1}} \cdots \Lambda_1 \xrightarrow{\partial_1} \Lambda_0$$

with the property that $\partial_{j-1} \circ \partial_j = 0$, $j = 2, \cdots, k$. The vector spaces Λ_j, $j = 0, 1 \cdots, k$, are called the spaces of j-chains. Each Λ_j is generated by a finite set of j-cells (like edges and vertices for graphs) in the sense that Λ_j is the set of functions from the j-cells to $\mathbb{R}$. A typical example of a k-complex is the triangularization of a k-dimensional manifold, with the j-cells, $j = 0, 1, \cdots, k$, being the sets of vertices, edges, faces, etc..

Example 2. Consider the triangularization of a 2-dimensional sphere by a tetrahedron with 4 faces, 6 edges, and 4 vertices. The matrix representation of the incidence operator ∂_2 (from the faces of the tetrahedron to its edges) is

	$< v_1 v_2 v_3 >$	$< v_1 v_3 v_4 >$	$< v_1 v_4 v_2 >$	$< v_2 v_4 v_3 >$
$< v_1 v_2 >$	1	0	-1	0
$< v_1 v_3 >$	-1	1	0	0
$< v_1 v_4 >$	0	-1	1	0
$< v_2 v_3 >$	1	0	0	-1
$< v_2 v_4 >$	0	0	-1	1
$< v_3 v_4 >$	0	1	0	-1

where the expressions $< v_1 v_2 v_3 >, \ldots$ denote the faces (with corresponding orientation), and $< v_1 v_2 >, \ldots$ are the edges.

---

[6] In algebraic topology [7] one usually starts with *abelian groups* Λ_j.

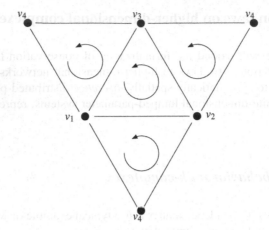

Fig. 1 Tetrahedron triangularizing a sphere

The matrix representation of the incidence operator ∂_1 (from edges to vertices) is given as

	$<v_1v_2>$	$<v_1v_3>$	$<v_1v_4>$	$<v_2v_3>$	$<v_2v_4>$	$<v_3v_4>$
$<v_1>$	-1	-1	-1	0	0	0
$<v_2>$	1	0	0	-1	-1	0
$<v_3>$	0	1	0	1	0	-1
$<v_4>$	0	0	1	0	1	1

It can be verified that $\partial_1 \circ \partial_2 = 0$.

Denoting the dual linear spaces by Λ^j, $j = 0, 1 \cdots, k$, we obtain the following dual sequence

$$\Lambda^0 \xrightarrow{d_1} \Lambda^1 \xrightarrow{d_2} \Lambda^2 \cdots \Lambda^{k-1} \xrightarrow{d_k} \Lambda^k$$

where the adjoint maps d_j $j = 0, 1 \cdots, k$, satisfy the analogous property $d_j \circ d_{j-1} = 0$, $j = 2, \cdots, k$. The elements of Λ^j are called j-cochains.

Consider any k-complex Λ, with k-chains $\alpha \in \Lambda_k$ and k-cochains $\beta \in \Lambda^k$. We define, similarly as in the case of a graph (1-complex) its Kirchhoff behavior as

$$B_K(\Lambda) := \{(\alpha, \beta) \in \Lambda_k \times \Lambda^k \mid$$
$$\partial_k \alpha = 0, \exists \phi \in \Lambda^{k-1} \text{ s.t. } \beta = d_k \phi\} \tag{19}$$

We will still refer to $\partial_k \alpha = 0$ as Kirchhoff's current laws (KCL), and to $\beta = d_k \psi$ as Kirchhoff's voltage laws (KVL). As before, it is immediately seen that $B_K(\Lambda) \subset \Lambda_k \times \Lambda^k$ is a *Dirac structure*. In particular, it follows that $< \beta \mid \alpha >_k = 0$ for every $(\alpha, \beta) \in B_K(\Lambda)$, where $< \cdot \mid \cdot >_k$ denotes the duality product between the dual linear spaces Λ_k and Λ^k.

3.2 Open k-complexes

Next we consider an *open* k-complex, by identifying a subset $\mathcal{V}^b_{(k-1)}$ of the set of all $(k-1)$-cells, called the *boundary* $(k-1)$-cells[7], while the remaining $(k-1)$-cells are denoted as the *internal* $(k-1)$-cells. Define the linear space of functions from this subset of $(k-1)$-cells to $\mathbb{R}$ as $\Lambda_b \subset \Lambda_{k-1}$ with dual space denoted as Λ^b. Decompose correspondingly $\partial_k : \Lambda_k \to \Lambda_{k-1}$ as $\partial_k = \begin{bmatrix} \partial^i_k \\ \partial^b_k \end{bmatrix}$, with adjoint mapping $d_k = d^i_k + d^b_k$. As before, Kirchhoff's voltage laws remain unchanged

$$\beta = d_k \psi = d^i_k \psi_i + d^b_k \psi_b, \tag{20}$$

where ψ_b is the vector of potentials at the boundary $(k-1)$-cells and ψ_i is the vector of potentials at the internal $(k-1)$-cells. On the other hand, Kirchhoff's current laws are modified to

$$\partial^i_k \alpha = 0, \quad \partial^b_k \alpha = -\alpha_b \tag{21}$$

where α_b denotes the vector of boundary 'currents' entering the boundary $(k-1)$-cells. By computing as before the total power we obtain for any α and β satisfying (20, 21)

$$< \beta \mid \alpha >_k = < d_k \psi \mid \alpha >_k = < d^i_k \psi_i + d^b_k \psi_b \mid \alpha >_k =$$
$$< \psi_i \mid \partial^i_k \alpha >_k + < \psi_b \mid \partial^b_k \alpha >_k = - < \psi_b \mid \alpha_b >_{k-1} \tag{22}$$

The space of boundary variables $(\alpha_b, \psi_b) \in \Lambda_b \times \Lambda^b$ describes the *distributed terminals* of the open k-complex.

Similar to Proposition 1 it is shown that the Kirchhoff behavior of an open k-complex Λ defined as

$$\mathcal{B}_K(\Lambda) := \{(\alpha, \beta, \alpha_b, \psi_b) \in \Lambda_k \times \Lambda^k \times \Lambda_b \times \Lambda^b \mid$$
$$\partial^i_k \alpha = 0, \partial^b_k = -\alpha_b, \exists \psi_i \text{ s.t. } \beta = d^i_k \psi_i + d^b_k \psi_b \} \tag{23}$$

is a Dirac structure.

Analogously to graphs, Kirchhoff current laws for open k-complexes imply certain constraints on the boundary 'currents' α_b. Indeed, by the fact that $\partial_{k-1} \circ \partial_k = 0$ it follows that $\partial_{(k-1)b} \alpha_b = 0$, where $\partial_{(k-1)b}$ denotes the $(k-1$-th incidence operator restricted to $\Lambda_b \subset \Lambda_{k-1}$. (Note that in the case of a graph the role of $\partial_{(k-1)b}$ is played by the linear map $\mathbb{1}^T_b$.) As in the case of graphs, this allows us to *reduce* the Kirchhoff behavior to a space that is still a Dirac structure, or, alternatively, to *close* the open k-complex. This is done by completing the open k-complex Λ with space of boundary currents Λ_b by an additional set of $(k-1)$-cells and k-cells.

---

[7] One could also consider as boundary cells subsets of the j-th cells for $j \neq k-1$. In particular, choosing $j = k$ would correspond to 'distributed interaction'. The choice $j = k-1$ corresponds to the important case of 'boundary' interaction.

Also the *interconnection* of open k-complexes is defined similar to the case of open graphs.

4 Port-Hamiltonian dynamics on k-complexes

Consider an open k-complex Λ, together with its Kirchhoff behavior $\mathscr{B}_K(\Lambda)$. Dynamics on the k-complex can be defined in various ways. Similar to the case of electrical circuits we could define constitutive relations for every k-cell, by specifying a relation between every component of Λ_k and Λ^k. As in the case of an electrical circuit this can be a relation of static resistive type, or a dynamic relation (of capacitive or inductive nature).

In this section we will define dynamics in a different way by specifying one type of dynamical relations between Λ_k and Λ^k, together with resistive relations between Λ_{k-1} and Λ^{k-1}. This will define a port-Hamiltonian dynamics, which is of *relaxation* type since there is only one type of physical energy (and thus no oscillations between different types of physical energy occur).

On the k-complex Λ, with $\partial_k : \Lambda_k \to \Lambda_{k-1}$ and $d_k : \Lambda^{k-1} \to \Lambda^k$, we define the following relations

$$f_x = -d_k e, \quad f_x \in \Lambda^k, e \in \Lambda^{k-1}$$

$$f = \partial_k e_x, \quad e_x \in \Lambda_k, f \in \Lambda_{k-1} \tag{24}$$

It is checked [16] that this defines a Dirac structure $\mathscr{D} \subset \Lambda^k \times \Lambda_k \times \Lambda^{k-1} \times \Lambda_{k-1}$. This allows us to define a port-Hamiltonian dynamics by imposing the following constitutive relations. First we associate to every k-cell an energy storage, leading to

$$\dot{x} = -f_x, \quad e_x = \frac{\partial H}{\partial x}(x), \quad x \in \Lambda^k \tag{25}$$

with $H(x)$ the total stored energy, and $x \in \Lambda^k$ the total vector of energy variables. Furthermore, we associate to every $(k-1)$-cell a (linear) resistive relation, leading to

$$e = -Rf, \quad R = R^T \geq 0 \tag{26}$$

Substituted in (24) this yields the relaxation dynamics

$$\dot{x} = d_k e = -d_k R f = -d_k R \partial_k \frac{\partial H}{\partial x}(x), \quad x \in \Lambda^k \tag{27}$$

with the property that

$$\frac{dH}{dt} = -(\partial_k \frac{\partial H}{\partial x}(x))^T R \partial_k \frac{\partial H}{\partial x}(x) = -f^T R f \leq 0 \tag{28}$$

For an *open* complex with boundary $(k-1)$-cells the definition is modified as follows. Instead of (24) we consider

$$f_x = -d_k \begin{bmatrix} e \\ e_b \end{bmatrix}, \quad f_x \in \Lambda^k, \begin{bmatrix} e \\ e_b \end{bmatrix} \in \Lambda^{k-1}, e_b \in \Lambda^b$$

$$\begin{bmatrix} f_i \\ f_b \end{bmatrix} = \partial_k e_x, \quad e_x \in \Lambda_k, \begin{bmatrix} f \\ f_b \end{bmatrix} \in \Lambda_{k-1}, f_b \in \Lambda_b$$

(29)

with f_b, e_b corresponding to the *boundary* $(k-1)$-cells, and f, e corresponding to the *internal* cells. Imposing the same storage relations (25) and resistive relations (26) we arrive at

$$\dot{x} = -d_k^r R \partial_k^r \frac{\partial H}{\partial x}(x) + d_k^b e_b$$

$$f_b = \partial_k^b \frac{\partial H}{\partial x}(x)$$

(30)

where we have split d_k as $d_k = \begin{bmatrix} d_k^r & d_k^b \end{bmatrix}$ and $\partial_k = \begin{bmatrix} \partial_k^r \\ \partial_k^b \end{bmatrix}$ (according to the division of the $(k-1)$-cells into internal cells corresponding to resistive behavior and boundary cells). This defines a port-Hamiltonian system with inputs e_b and outputs f_b.

4.1 Example: Heat transfer on a 2-complex

The above formulation of systems of conservation laws and port-Hamiltonian systems on k-complexes will be illustrated with the model of heat transfer in a 2-dimensional medium (for instance a plate). Instead of first considering the pde-model and then discretizing, we will directly consider the dynamics on a 2-complex as arising from a triangulation of the 2-dimensional spatial domain. We assume the medium to be undeformable (hence mechanical work is neglected) and that there is no mass transfer.

We will write the heat transfer in terms of the conservation of internal energy. First we identify the physical variables as chains and cochains of the given 2-complex. The components of the internal energy vector $u \in \Lambda^2$ denote the energy of each face. The heat conduction is given by the *heat flux* $f \in \Lambda^1$ whose components equal the heat flux through every edge. Hence the basic conservation law (conservation of energy) is given as

$$\frac{du}{dt} = d_2 f$$

The thermodynamic properties are defined by Gibbs' relation, and generated by the *entropy function* $s = s(u) \in C^\infty(\Lambda^2)$ as thermodynamic potential. Since we consider transformations which are isochore and without mass transfer, Gibbs' relation reduces to the definition of the vector of intensive variables $e_u \in \Lambda_2$ which is (entropy-)conjugated to the vector of extensive variables u:

$$e_u = \frac{\partial s}{\partial u}(u)$$

The components e_u are equal to the reciprocal of the temperature at each 2-face.

Since the temperature is varying over the faces, there is a *thermodynamic driving force* vector $e \in \Lambda_1$ given as the vector of differences

$$e = \partial_2 e_u$$

By Fourier's law the heat flux is determined by the thermodynamic driving force vector as

$$f = R(e_u)e, \tag{31}$$

with $R(e_u) = R^T(e_u) \geq 0$ depending on the heat conduction coefficients. (Note the sign-difference with (26).) The resulting system is a port-Hamiltonian system (of relaxation type), with vector of state variables x given by the internal energy vector u, and 'Hamiltonian' $s(u)$. By (28) the entropy $s(u)$ satisfies

$$\frac{ds}{dt} = (\partial_2 \frac{\partial s}{\partial u}(u))^T R(e_u) \partial_2 \frac{\partial s}{\partial u}(u) = f^T R(e_u) f \geq 0$$

expressing the fact that the entropy is monotonously *increasing*. (Note again the sign-difference with the treatment above, where the Hamiltonian H was decreasing.)

The exchange of heat through the boundary of the system is incorporated as above, cf. (29, 30), by splitting the edges (1-cells) into internal edges with the resistive relation (31) and boundary edges. This leads to

$$\frac{ds}{dt} = (\partial_2 \frac{\partial s}{\partial u}(u))^T R(e_u) \partial_2 \frac{\partial s}{\partial u}(u) + e_b f_b$$

with f_b, e_b denoting the heat flux, respectively, thermodynamical driving force, through the boundary edges.

5 Conclusions

A framework has been laid down for the formulation of open physical systems on k-complexes, generalizing the graph-theoretic formulation of electrical circuit dynamics with terminals. It has been shown that Kirchhoff's laws can be generalized to open k-complexes, defining a Dirac structure involving boundary currents and potentials, thus generalizing the concept of 'terminal' to the distributed case. This has been illustrated on the example of heat transfer on a 2-complex. This simple example already shows how one can directly define a finite-dimensional port-Hamiltonian dynamics, capturing the physical meaning of the involved variables and retaining the conservation laws, without the need to formulate the dynamics as a set of pde's (and possibly to discretize the pde's later on).

In future work we will apply and extend the framework to different classes of port-Hamiltonian systems on k-complexes (corresponding to different physical settings), and employ these models for boundary control.

Acknowledgements It is a great pleasure to contribute to this *Liber Amicorum* for Professor Okko Bosgra at the occasion of his 65-th birthday. Okko has played a key role in the flourishing of the Systems & Control community within the Netherlands, both by his scientific leadership at the Delft University of Technology, as well as by his stimulating role in bringing control engineering and mathematical systems and control theory together. I hope this chapter pays tribute to his pioneering achievements and visionary contributions.

References

[1] Bamberg, P., Sternberg, S.: A course in mathematics for students of physics: 2. Cambridge University Press (1999)

[2] Bird, R.B., Stewart, W.E., Lightfoot, E.N.: Transport Phenomena. John Wiley and Sons (2002)

[3] Bollobas, B.: Modern Graph Theory. Graduate Texts in Mathematics, vol. 184, Springer, New York (1998)

[4] Bosgra, O.H.: Lecture Notes 'Mathematical Modelling of Dynamical Engineering Systems'. Delft University of Technology.

[5] Cervera, J., van der Schaft, A.J., Banos, A.: Interconnection of port-Hamiltonian systems and composition of Dirac structures. Automatica **43**, 212–225 (2007)

[6] Courant, T.J.: Dirac manifolds. Trans. Amer. Math. Soc. **319**, 631–661 (1990)

[7] Dold, A.: Lectures on Algebraic Topology. Classics in Mathematics, Springer, Berlin (1995)

[8] Golo, G., Talasila, V., van der Schaft, A.J., Maschke, B.M.: Hamiltonian discretization of boundary control systems. Automatica **40**(5), 757–711 (2004)

[9] Kirchhoff, G.: Über die Auflösung der Gleichungen, auf welche man bei der Untersuchung der Linearen Verteilung galvanischer Ströme geführt wird. Ann. Phys. Chem., vol. 72, 497–508 (1847)

[10] Kondo, K., Iri, M.: On the theory of trees, cotrees, multi-trees and multi-cotrees. RAAG Memoirs **2**, Div. A-VII, 220–261 (1958)

[11] MacFarlane, A.G.J.: Dynamical System Models. G.G. Harrap & Co., London (1970)

[12] Mizoo, Y., Iri, M., Kondo, K.: On the torsion and linkage characteristics and the duality of electric, magnetic and dielectric networks. RAAG Memoirs **2**, Div. A-VIII, 262-295 (1958)

[13] van der Schaft, A.J.: L_2-Gain and Passivity Techniques in Nonlinear Control. Lect. Notes in Control and Information Sciences, vol. 218, Springer-Verlag, Berlin, (1996), p. 168, 2nd revised and enlarged edition, Springer-Verlag, London (2000) (Springer Communications and Control Engineering series), p. xvi+249.

[14] van der Schaft, A.J., Maschke, B.M.: The Hamiltonian formulation of energy conserving physical systems with external ports. Archiv für Elektronik und Übertragungstechnik **49**, 362–371 (1995)

[15] van der Schaft, A.J., Maschke, B.M.: Hamiltonian formulation of distributed-parameter systems with boundary energy flow. Journal of Geometry and Physics **42**, 166–194 (2002)

[16] van der Schaft, A.J., Maschke, B.M.: Conservation laws and open systems on higher-dimensional networks. In: Proc. 47th IEEE Conf. on Decision and Control, Cancun, Mexico, December 9-11, 799–804 (2008)

[17] Willems, J.C.: The behavioral approach to open and interconnected systems. Control Systems Magazine **27**, Dec., 46–99 (2007)

Polynomial Optimization Problems are Eigenvalue Problems

Philippe Dreesen and Bart De Moor

To our good friend and colleague Okko Bosgra
For the many scientific interactions, the wise and thoughtful advise at many occasions
The stimulating and pleasant social interactions
Okko, ad multos annos!

Abstract Many problems encountered in systems theory and system identification require the solution of polynomial optimization problems, which have a polynomial objective function and polynomial constraints. Applying the method of Lagrange multipliers yields a set of multivariate polynomial equations. Solving a set of multivariate polynomials is an old, yet very relevant problem. It is little known that behind the scene, linear algebra and realization theory play a crucial role in understanding this problem. We show that determining the number of roots is essentially a linear algebra question, from which we derive the inspiration to develop a root-finding algorithm based on realization theory, using eigenvalue problems. Moreover, since one is only interested in the root that minimizes the objective function, power iterations can be used to obtain the minimizing root directly. We highlight applications in systems theory and system identification, such as analyzing the convergence behaviour of prediction error methods and solving structured total least squares problems.

1 Introduction

Solving a polynomial optimization problem is a core problem in many scientific and engineering applications. Polynomial optimization problems are numerical optimization problems where both the objective function and the constraints are multivariate polynomials. Applying the method of Lagrange multipliers yields necessary conditions for optimality, which, in the case of polynomial optimization, results in a set of multivariate polynomial equations.

Katholieke Universiteit Leuven, Department of Electrical Engineering – ESAT/SCD, Kasteelpark Arenberg 10, 3001 Leuven, Belgium. e-mail: philippe.dreesen,bart.demoor@esat.kuleuven.be

P.M.J. Van den Hof et al. (eds.), *Model-Based Control: Bridging Rigorous Theory and Advanced Technology*, DOI: 10.1007/978-1-4419-0895-7_4, © Springer Science + Business Media, LLC 2009

Solving a set of multivariate polynomial equations, i.e., calculating its roots, is an old problem, which has been studied extensively. For a brief history and a survey of relevant notions regarding both algebraic geometry and optimization theory, we refer to the excellent text books [5, 6] and [35], respectively. The technique presented in this contribution explores formulations based on the Sylvester and Macaulay constructions [27, 28, 42], which 'linearize' a set of polynomial equations in terms of linear algebra. The resulting task can then be tackled using well-understood matrix computations, such as eigenvalue calculations. We present an algorithm, based on realization theory, to obtain all solutions of the original set of equations from a (generalized) eigenvalue problem. Since in most cases, one is only interested in the root that minimizes the given objective function, the global minimizer can be determined directly as a maximal eigenvalue. This can be achieved using well-understood matrix computations such as power iterations [15]. Applications are ubiquitous, and include solving and analyzing the convergence behaviour of prediction error system identification methods [24], and solving structured total least squares problems [7, 8, 9, 10, 11].

This contribution comprises a collection of ideas and research questions. We aim at revealing and exploring vital connections between systems theory and system identification, optimization theory, and algebraic geometry. For convenience, we assume that the coefficients of the polynomials used are real, and we also assume that the sets of polynomial equations encountered in our examples describe zero-dimensional varieties (the set of polynomial equations has a finite number of solutions). In order to clarify ideas, we limit the mathematical content, and illustrate the presented methods with several simple instructional examples. The purpose of this paper is to share some interesting ideas. The real challenges lie in deriving the proofs, and moreover in the development of efficient algorithms to overcome numerical and combinatorial issues.

The remainder of this paper is organized as follows: Section 2 gives the outline of the proposed technique; We derive a rank test that counts the number of roots, and a method to retrieve the solutions. Furthermore, we propose a technique that allows to identify the minimizing root of a polynomial cost criterion, and we briefly survey the most important current methods for solving systems of multivariate polynomial equations. In Section 3, a list of relevant applications is outlined. Finally, Section 4 provides the conclusions and describes the main research problems and future work.

2 General Theory

2.1 Introduction

In this section, we lay out a path from polynomial optimization problems to eigenvalue problems. The main steps are phrasing the task at hand as a set of polynomial equations by applying the method of Lagrange multipliers, casting the problem of solving a set of polynomial equations into an eigenvalue problem, and applying matrix algebra methods to solve it.

2.2 Polynomial Optimization is Polynomial System Solving

Consider a polynomial optimization problem

$$\min_{\mathbf{x} \in \mathbb{C}^p} \quad J(\mathbf{x}) \tag{1}$$

$$\text{s. t.} \quad g_i(\mathbf{x}) = 0, \quad i = 1, \dots, q,$$

where $J : \mathbb{C}^p \to \mathbb{R}$ is the polynomial objective function, and $g_i : \mathbb{C}^p \to \mathbb{R}$ represent q polynomial equality constraints in the unknowns $\mathbf{x} = (x_1, \dots, x_p)^T \in \mathbb{C}^p$. We assume that all coefficients in $J(\cdot)$ and $g_i(\cdot)$ are real. The Lagrange multipliers method yields the necessary first-order conditions for optimality: they are found from the stationary points of the Lagrangian function, for which we introduce a vector $\mathbf{a} = (\alpha_1, \dots, \alpha_q)^T \in \mathbb{R}^q$, containing the Lagrange multipliers α_i:

$$\mathcal{L}(\mathbf{x}, \mathbf{a}) = J(\mathbf{x}) + \sum_{i=1}^{p} \alpha_i g_i(\mathbf{x}). \tag{2}$$

This results in

$$\frac{\partial}{\partial x_i} \mathcal{L}(\mathbf{x}, \mathbf{a}) = 0, \qquad i = 1, \dots, p, \quad \text{and} \tag{3}$$

$$\frac{\partial}{\partial \alpha_i} \mathcal{L}(\mathbf{x}, \mathbf{a}) = 0, \qquad i = 1, \dots, q. \tag{4}$$

In the case of a polynomial optimization problem, it is easy to see that the result is a set of $m = p + q$ polynomial equations in $n = p + q$ unknowns.

Example 1. Consider

$$\min_x \quad J = x^2 \tag{5}$$

$$\text{s. t.} \quad (x - 1)(x - 2) = 0. \tag{6}$$

Since the constraint has only two solutions ($x = 1$ and $x = 2$), it is easily verified that the solution of this problem corresponds to $x = 1$, for which the value of the objective function J is 1. In general, this type of problem can be tackled using Lagrange multipliers. The Lagrangian function is given by

$$\mathcal{L}(x, \alpha) = x^2 + \alpha(x - 1)(x - 2), \tag{7}$$

where α denotes a Lagrange multiplier. We find the necessary conditions for optimality from the stationary points of the Lagrangian function as a set of two polynomial equations in the unknowns x and α:

$$\frac{\partial \mathscr{L}}{\partial x}(x,\alpha) = 0 \qquad \Longleftrightarrow \qquad 2x + 2x\alpha - 3\alpha = 0, \tag{8}$$

$$\frac{\partial \mathscr{L}}{\partial \alpha}(x,\alpha) = 0 \qquad \Longleftrightarrow \qquad x^2 - 3x + 2 = 0. \tag{9}$$

In this example, the stationary points (x,α) are indeed found as $(1,2)$ and $(2,-4)$. The minimizing solution of the optimization problem is found as $(x,\alpha) = (1,2)$.

The conclusion of this section is that solving polynomial optimization problems results in finding roots of sets of multivariate polynomial equations.

2.3 Solving a System of Polynomial Equations is Linear Algebra

2.3.1 Motivational Example

It is little known that behind the scenes, the task of solving a set of polynomial equations is essentially a linear algebra question. In order to illustrate this, a motivational example is introduced to which we will return throughout the remainder of this paper.

Example 2. Consider a simple set of two equations in x and y, borrowed from [41]:

$$p_1(x,y) = 3 + 2y - xy - y^2 + x^2 = 0 \tag{10}$$
$$p_2(x,y) = 5 + 4y + 3x + xy - 2y^2 + x^2 = 0. \tag{11}$$

This system has four solutions for (x,y); two real solutions and a complex conjugated pair: $(0.08, 2.93)$, $(-4.53, -2.58)$, and $(0.12 \mp 0.70i, -0.87 \pm 0.22i)$. The system is visualized in Fig. 1.

2.3.2 Preliminary Notions

We start by constructing bases of monomials. Let n denote the number of unknowns. In accordance with example Eq. (10)–(11), we consider the case of $n = 2$ unknowns x and y.

Let $\mathbf{w}_\delta$ be a basis of monomials of degree δ in two unknowns x and y, constructed as

$$\mathbf{w}_\delta = (x^\delta, x^{\delta-1}y, \ldots, y^\delta)^T. \tag{12}$$

Given a maximum degree d, a column vector containing a full basis of monomials $\mathbf{v}_d$ is constructed by stacking bases $\mathbf{w}_\delta$ of degrees $\delta \leq d$:

$$\mathbf{v}_d = (\mathbf{w}_0; \mathbf{w}_1; \ldots; \mathbf{w}_d). \tag{13}$$

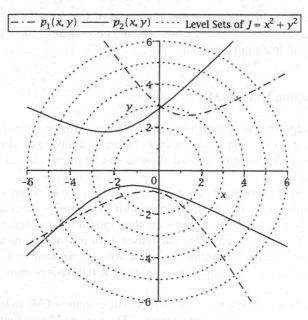

Fig. 1 A simple set of two polynomials $p_1(x,y) = 0$ and $p_2(x,y) = 0$, see Eq. (10) (dot-dashed line) and Eq. (11) (solid line), respectively. This system has four (complex) solutions (x,y): $(0.08, 2.93)$, $(-4.53, -2.58)$, and $(0.12 \mp 0.70i, -0.87 \pm 0.22i)$. Level sets of a polynomial objective function $J = x^2 + y^2$ are also shown (dotted lines).

Example 3. For the case of two unknowns x and y, and $\delta \leq 3$, we have

$$
\begin{aligned}
\mathbf{w}_0 &= (1)^T, & \mathbf{v}_0 &= (1)^T, \\
\mathbf{w}_1 &= (x, y)^T, & \mathbf{v}_1 &= (1, x, y)^T, \\
\mathbf{w}_2 &= (x^2, xy, y^2)^T, & \mathbf{v}_2 &= (1, x, y, x^2, xy, y^2)^T, \\
\mathbf{w}_3 &= (x^3, x^2y, xy^2, y^3)^T, & \mathbf{v}_3 &= (1, x, y, x^2, xy, y^2, x^3, x^2y, xy^2, y^3)^T.
\end{aligned}
\tag{14}
$$

Observe the 'shift properties' inherent in this formulation, which will turn out to play a fundamental role in our root-counting technique and root-finding algorithm. By shift properties, we mean that by multiplication with a certain monomial, e.g., x or y, monomials of low degree are mapped to monomials of higher degree. For example:

$$
\begin{aligned}
(1, x, y, x^2, xy, y^2)^T x &= (x, x^2, xy, x^3, x^2y, xy^2)^T, \\
(1, x, y, x^2, xy, y^2)^T y &= (y, xy, y^2, x^2y, xy^2, y^3)^T.
\end{aligned}
\tag{15}
$$

Let us end with some remarks. Firstly, as the number of unknowns used should be clear in each example, we have dismissed the explicit dependence of $\mathbf{v}_d$ and $\mathbf{w}_\delta$ on n for notational convenience. Secondly, we have used a specific monomial order, i.e., graded lexicographic order: monomials are first ordered by their total degree, terms of equal total degree are then ordered lexicographically. The techniques illustrated here can be easily generalized to other monomial orders. Thirdly, we have only discussed the case of $n = 2$ unknowns, however, the general case of $n > 2$ unknowns can be worked out in a similar fashion.

2.3.3 Constructing Matrices $\mathbf{M}_d$

We outline an algorithm to construct so-called Macaulay-like matrices $\mathbf{M}_d$, satisfying $\mathbf{M}_d \mathbf{v}_d = 0$, where $\mathbf{v}_d$ is a column vector containing a full basis of monomials, as defined above. This construction will allow us to cast the problem at hand into a linear algebra question.

Given a set of m polynomial equations p_i, $i = 1, 2, \ldots, m$ of degrees d_i in n unknowns. We will assume that the number of multivariate polynomial equations m equals the number of unknowns n, as this is the case for all the examples we will discuss in this paper. Also recall that only cases with few unknowns are encountered, hence, for notational convenience, unknowns are denoted by x, y, etc. Let $d° = \max(d_i)$ and let $d \geq d°$ denote the degree of the full basis of monomials $\mathbf{v}_d$ to be used.

The construction of $\mathbf{M}_d$ proceeds as follows: The columns of $\mathbf{M}_d$ index monomials of degrees d or less, i.e., the elements of $\mathbf{v}_d$. The rows of $\mathbf{M}_d$ are found from the forms $r \cdot p_i$, where r is a monomial for which $\deg(r \cdot p_i) \leq d$. The entries of a row are hence found by writing the coefficients of the corresponding forms $r \cdot p_i$ in the appropriate columns. In this way, each row of $\mathbf{M}_d$ corresponds to one of the input polynomials p_i.

Example 4. This process is illustrated using Eq. (10)–(11). The 2×6 matrix $\mathbf{M}_2$ is found from the original set of equations directly as

$$
\mathbf{M}_2 \mathbf{v}_2 = 0 \quad \text{or} \quad \begin{pmatrix} 3 & 0 & 2 & 1 & -1 & -1 \\ 5 & 3 & 4 & 1 & 1 & -2 \end{pmatrix} \begin{pmatrix} 1 \\ x \\ y \\ x^2 \\ xy \\ y^2 \end{pmatrix} = 0. \tag{16}
$$

We now increase d to $d = 3$, and add rows to complete $\mathbf{M}_3$, yielding a 6×10 matrix. By increasing the degree d to 4, we obtain a 12×15 matrix $\mathbf{M}_4$. This process is shown in Tab. 1.

In general, the matrices generated in this way are very sparse, and typically quasi-Toeplitz structured [32]. As d is increased further, the number of monomials in the corresponding bases of monomials $\mathbf{v}_d$ of the matrices $\mathbf{M}_d$ grows. The number of monomials of given degree d in n unknowns is given by the relation

$$
N(n,d) = \frac{(d+n-1)!}{d!\,(n-1)!}. \tag{17}
$$

Table 1 The construction of matrices $\mathbf{M}_d$. Columns of $\mathbf{M}_d$ are indexed by monomials of degrees d and less. Rows of $\mathbf{M}_d$ are found from the forms $r \cdot p_i$, where r is a monomial and $\deg(r \cdot p_i) \leq d$; the entries of each row are found by writing the coefficients of the input polynomials in the appropriate columns. The construction process is shown for $d = 2$ (resulting in a 2×6 matrix), up to $d = 4$ (resulting in a 12×15 matrix). Empty spaces represent zero elements.

		1	x	y	x^2	xy	y^2	x^3	x^2y	xy^2	y^3	x^4	x^3y	x^2y^2	xy^3	y^4	...
$d° = 2$	p_1	3	2		1	-1	-1										
	p_2	5	3	4	1	1	-2										
$d = 3$	$x \cdot p_1$		3		2			1	-1	-1							
	$y \cdot p_1$			3		2			1	-1	-1						
	$x \cdot p_2$		5		3	4		1	1	-2							
	$y \cdot p_2$			5		3	4		1	1	-2						
$d = 4$	$x^2 \cdot p_1$				3			2				1	-1	-1			
	$xy \cdot p_1$					3			2				1	-1	-1		
	$y^2 \cdot p_1$						3			2				1	-1	-1	
	$x^2 \cdot p_2$				5			3	4			1	1	-2			
	$xy \cdot p_2$					5			3	4			1	1	-2		
	$y^2 \cdot p_2$						5			3	4			1	1	-2	

As d increases, the row-size (number of forms $r \cdot p_i$) increases too. It can be verified that the number of rows in $\mathbf{M}_d$ grows faster than the number of monomials in $\mathbf{v}_d$, since each input polynomial is multiplied by a set of monomials, the number of which increases according to Eq. (17). Therefore, a degree d^* exists, for which the corresponding matrix $\mathbf{M}_{d^*}$ has more rows than columns. It turns out that the number of solutions, and, moreover, the solutions themselves, can be retrieved from the 'overdetermined' matrix $\mathbf{M}_{d^*}$, as will be illustrated in the following sections.

For the example of Eq. (10)–(11), we find $d^* = 6$, and a corresponding 30×28 matrix $\mathbf{M}_6$. Due to the relatively large size, this matrix is not printed, however, its construction is straightforward, and proceeds as illustrated in Tab. 1.

Observe that in Eq. (10)–(11), the given polynomials are both of degree two. If the input polynomials are not of the same degree, one also has to shift the input polynomial(s) of lower degree internally: For example, consider the case of two (univariate) polynomials

$$a_3 x^3 + a_2 x^2 + a_1 x + a_0 = 0 \tag{18}$$
$$b_2 x^2 + b_1 x + b_0 = 0, \tag{19}$$

then, for $d = 3$, one easily finds

$$\begin{pmatrix} a_0 & a_1 & a_2 & a_3 \\ b_0 & b_1 & b_2 & 0 \\ 0 & b_0 & b_1 & b_2 \end{pmatrix} \begin{pmatrix} 1 \\ x \\ x^2 \\ x^3 \end{pmatrix} = 0, \tag{20}$$

where the third row in the matrix is found by multiplying the second polynomial with x. Increasing d to 4 yields the classical 5×5 Sylvester matrix.

2.4 Determining the Number of Roots

From the construction we just described, it is easy to verify that the constructed matrices $\mathbf{M}_d$ are rank-deficient: by evaluating the basis of monomials $\mathbf{v}_d$ at the (for the moment unknown) roots, a solution for $\mathbf{M}_d\mathbf{v}_d = 0$ is found. By construction, in the null space of the matrices $\mathbf{M}_d$, we find vectors $\mathbf{v}_d$ containing the roots of the set of multivariate polynomial equations. As a matter of fact, every solution generates a different vector $\mathbf{v}_d$ in the null space of $\mathbf{M}_d$. Obviously, when $\mathbf{M}_d$ is overdetermined, it must become rank deficient because of the vectors $\mathbf{v}_d$ in the kernel of $\mathbf{M}_d$. Contrary to what one might think, the co-rank of $\mathbf{M}_d$, i.e., the dimension of the null space, does not provide the number of solutions directly; It represents an upper bound. The reason for this is that certain vectors in the null space are 'spurious': they do not contain information on the roots. Based on these notions, we will now work out a root-counting technique: we reason that the exact number of solutions derives from the notion of a *truncated co-rank*, as will be illustrated in this section.

Observe in Tab. 1 that the matrices $\mathbf{M}_d$ of low degrees are embedded in those of high degree, e.g., $\mathbf{M}_2$ occurs as the upper-left block in an appropriately partitioned $\mathbf{M}_3$, $\mathbf{M}_3$ in $\mathbf{M}_4$, etc. For Eq. (10)–(11), one has the following situation:

$$
\begin{array}{c||cccccccc||c}
d & \mathbf{w}_0 & \mathbf{w}_1 & \mathbf{w}_2 & \mathbf{w}_3 & \mathbf{w}_4 & \mathbf{w}_5 & \mathbf{w}_6 & \mathbf{w}_7 & \text{size } \mathbf{M}_d \\
\hline
2(=d^\circ) & \times & \times & \times & 0 & 0 & 0 & 0 & 0 & 2 \times 6 \\
3 & 0 & \times & \times & \times & 0 & 0 & 0 & 0 & 6 \times 10 \\
4 & 0 & 0 & \times & \times & \times & 0 & 0 & 0 & 12 \times 15 \\
5 & 0 & 0 & 0 & \times & \times & \times & 0 & 0 & 20 \times 21 \\
6(=d^\star) & 0 & 0 & 0 & 0 & \times & \times & \times & 0 & \mathbf{30 \times 28} \\
7 & 0 & 0 & 0 & 0 & 0 & \times & \times & \times & 42 \times 36
\end{array}
\tag{21}
$$

In the left-most column, the degree d is indicated. The right-most column shows the size of the corresponding matrix $\mathbf{M}_d$. The middle part represents the matrix $\mathbf{M}_d$. The columns of $\mathbf{M}_d$ are indexed (block-wise) by bases of monomials $\mathbf{w}_\delta$. Zeros in $\mathbf{M}_d$ represent zero-blocks, and blocks of non-zero entries are indicated by $\times$.

For Eq. (10)–(11) we have found that for $d^\star = 6$, the matrix $\mathbf{M}_{d^\star}$ becomes overdetermined. Let us now investigate what happens to the matrices $\mathbf{M}_d$ when increasing d. Consider the transition from, say, $d = 6$ to $d = 7$. We partition $\mathbf{M}_6$ as follows (cf. Eq. (21)):

$$
\mathbf{M}_6 = \begin{pmatrix}
\times & \times & \times & 0 & 0 & 0 & 0 \\
0 & \times & \times & \times & 0 & 0 & 0 \\
0 & 0 & \times & \times & \times & 0 & 0 \\
0 & 0 & 0 & \times & \times & \times & 0 \\
0 & 0 & 0 & 0 & \times & \times & \times
\end{pmatrix} = (\,\mathbf{K}_6 \,|\, \mathbf{L}_6\,).
\tag{22}
$$

where the number of columns in $\mathbf{K}_6$ corresponds to the number of monomials in the bases of monomials $\mathbf{w}_0$ to $\mathbf{w}_4$. Since $\mathbf{M}_6$ is embedded in $\mathbf{M}_7$, we can also identify $\mathbf{K}_6$ and $\mathbf{L}_6$ in $\mathbf{M}_7$. We partition $\mathbf{M}_7$ accordingly:

$$
\mathbf{M}_7 = \begin{pmatrix}
\times & \times & \times & 0 & 0 & 0 & 0 & 0 \\
0 & \times & \times & \times & 0 & 0 & 0 & 0 \\
0 & 0 & \times & \times & \times & 0 & 0 & 0 \\
0 & 0 & 0 & \times & \times & \times & 0 & 0 \\
0 & 0 & 0 & 0 & \times & \times & \times & 0 \\
0 & 0 & 0 & 0 & 0 & \times & \times & \times
\end{pmatrix} = \left(\begin{array}{c|c|c} \mathbf{K}_6 & \mathbf{L}_6 & 0 \\ \hline 0 & \mathbf{D}_7 & \mathbf{E}_7 \end{array} \right)
\tag{23}
$$

Let $\mathbf{Z}_6$ and $\mathbf{Z}_7$ denote (numerically computed) kernels of $\mathbf{M}_6$ and $\mathbf{M}_7$, respectively. The kernels $\mathbf{Z}_6$ and $\mathbf{Z}_7$ are now partitioned accordingly, such that:

$$
\mathbf{M}_6\mathbf{Z}_6 = \left(\mathbf{K}_6 \,|\, \mathbf{L}_6 \right) \left(\frac{\mathbf{T}_6}{\mathbf{B}_6} \right), \qquad \text{and} \tag{24}
$$

$$
\mathbf{M}_7\mathbf{Z}_7 = \left(\begin{array}{c|c} \mathbf{K}_6 & \mathbf{G}_7 \\ \hline 0 & \mathbf{H}_7 \end{array} \right) \left(\frac{\mathbf{T}_7}{\mathbf{B}_7} \right), \tag{25}
$$

where

$$
\mathbf{G}_7 = \left(\mathbf{L}_6 \,|\, 0 \right), \qquad \text{and} \tag{26}
$$
$$
\mathbf{H}_7 = \left(\mathbf{D}_7 \,|\, \mathbf{E}_7 \right). \tag{27}
$$

Due to the zero block in $\mathbf{M}_7$ below $\mathbf{K}_6$, we have

$$
\text{rank}(\mathbf{T}_6) = \text{rank}(\mathbf{T}_7), \tag{28}
$$

which we will call the *truncated co-rank* of $\mathbf{M}_{d^\star}$. In other words, for $d \geq d^\star$, the rank of the appropriately truncated bases for the null spaces, stabilizes. We call the rank at which it stabilizes the truncated co-rank of $\mathbf{M}_d$. It turns out that the number of solutions corresponds to the truncated co-rank of $\mathbf{M}_d$, given that a sufficiently high degree d is chosen, i.e., $d \geq d^\star$. Correspondingly, we will call the matrices $\mathbf{T}_6$ and $\mathbf{T}_7$ the *truncated kernels*, from which the solutions can be retrieved, as explained in the following section.

2.5 Finding the Roots

The results from the previous section allow us to find the number of roots as a rank test on the matrices $\mathbf{M}_d$ and their truncated kernels. We will now present a root-finding algorithm, inspired by realization theory and the shift property Eq. (15), which reduces to the solution of a generalized eigenvalue problem. We will also point out other approaches to phrase the task at hand as an eigenvalue problem.

2.5.1 Realization Theory

We will illustrate the approach using the example Eq. (10)–(11).

Example 5. A matrix $\mathbf{M}_d$ of sufficiently high degree was constructed in Section 2.3.3: we found $d^\star = 6$ and constructed $\mathbf{M}_6$. One can verify that the truncated co-rank of $\mathbf{M}_6$ is four, hence there are four solutions. Note that one can always construct a *canonical form* of the (truncated) kernel $\mathbf{V}_d$ of $\mathbf{M}_d$ as follows, e.g., for $d = 6$ and four different roots:

$$
\mathbf{V}_6 = \left(\begin{array}{c|c|c|c}
1 & 1 & 1 & 1 \\
\hline
x_1 & x_2 & x_3 & x_4 \\
y_1 & y_2 & y_3 & y_4 \\
x_1^2 & x_2^2 & x_3^2 & x_4^2 \\
x_1 y_1 & x_2 y_2 & x_3 y_3 & x_4 y_4 \\
y_1^2 & y_2^2 & y_3^2 & y_4^2 \\
x_1^3 & x_2^3 & x_3^3 & x_4^3 \\
x_1^2 y_1 & x_2^2 y_2 & x_3^2 y_3 & x_4^2 y_4 \\
x_1 y_1^2 & x_2 y_2^2 & x_3 y_3^2 & x_4 y_4^2 \\
y_1^3 & y_2^3 & y_3^3 & y_4^3 \\
\vdots & \vdots & \vdots & \vdots
\end{array}\right),
\tag{29}
$$

where (x_i, y_i), $i = 1, \ldots, 4$ represent the four roots of Eq. (10)–(11). Note that the truncated kernels are used to retrieve the roots, this means that certain rows are omitted from the full kernels, in accordance with the specific partitioning discussed in Section 2.4. Observe that the columns of this generalized Vandermonde matrix $\mathbf{V}_d$ are nothing more than all roots evaluated in the monomials indexing the columns of $\mathbf{M}_d$. Let us recall the shift property Eq. (15): if we multiply the upper part of one of the columns of $\mathbf{V}_6$ with x, we have:

$$
\begin{pmatrix} 1 \\ \hline x \\ y \\ \hline x^2 \\ xy \\ y^2 \end{pmatrix} x = \begin{pmatrix} x \\ x^2 \\ xy \\ x^3 \\ x^2 y \\ xy^2 \end{pmatrix}.
\tag{30}
$$

Let $\mathbf{D} = \operatorname{diag}(x_1, x_2, x_3, x_4)$ be a diagonal matrix with the (for the moment unknown) x-roots. In accordance with the shift property Eq. (15), we can write

$$
\mathbf{S}_1 \mathbf{V}_6 \mathbf{D} = \mathbf{S}_2 \mathbf{V}_6,
\tag{31}
$$

where $\mathbf{S}_1$ and $\mathbf{S}_2$ are so-called row selection matrices: $\mathbf{S}_1 \mathbf{V}_6$ selects the first rows of $\mathbf{V}_6$, corresponding to degrees 1 to $5 (= 6 - 1)$. $\mathbf{S}_2 \mathbf{V}_6$ represents rows 2, 4, 5, 7, 8, 9, etc. of $\mathbf{V}_6$, in order to perform the multiplication with x. In general, the kernel $\mathbf{V}_d$ is not available in the *canonical form* as in Eq. (3). Instead, a kernel $\mathbf{Z}_d$ is calculated numerically as

$$\mathbf{M}_d \mathbf{Z}_d = 0, \tag{32}$$

where $\mathbf{Z}_d = \mathbf{V}_d \mathbf{T}$, for a certain non-singular matrix $\mathbf{T}$. Hence,

$$\mathbf{S}_1 \mathbf{Z}_d (\mathbf{T}^{-1} \mathbf{D} \mathbf{T}) = \mathbf{S}_2 \mathbf{Z}_d, \tag{33}$$

and the root-finding problem is reduced to a generalized eigenvalue problem:

$$(\mathbf{Z}_d^T \mathbf{S}_1^T \mathbf{S}_2 \mathbf{Z}_d) \mathbf{u} = \lambda (\mathbf{Z}_d^T \mathbf{S}_1^T \mathbf{S}_1 \mathbf{Z}_d) \mathbf{u}. \tag{34}$$

We revisit example Eq. (10)–(11). A kernel of $\mathbf{M}_6$, i.e., $\mathbf{Z}_6$, is computed numerically using a singular value decomposition. The number of solutions was obtained in the previous sections as four. After constructing appropriate selection matrices $\mathbf{S}_1$ and $\mathbf{S}_2$, and solving the resulting generalized eigenvalue problem, the roots (x, y) are found as in Example 2.

2.5.2 The Stetter-Möller Eigenvalue Problem

It is a well-known fact that the roots of a univariate polynomial correspond to the eigenvalues of the corresponding Frobenius companion matrix (this is how the roots are computed using the **roots** command in MATLAB). The notion that a set of multivariate polynomial equations can also be reduced to an eigenvalue problem was known to Sylvester and Macaulay already in the late 19th and early 20th century, but was only recently rediscovered by Stetter and coworkers (cf. [1, 31, 39], similar approaches are [19, 21, 30]). We will recall the main ideas from this framework, and illustrate the technique using example Eq. (10)–(11).

In order to phrase the problem of solving a set of multivariate polynomial equations as an eigenvalue problem, one needs to construct a monomial basis, prove that it is *closed* for multiplication with the unknowns for which one searches the roots. Furthermore, one needs to construct associated multiplication matrices (cf. [39, 40]). In practice, this can be accomplished easily after applying a normal form algorithm, such as the Gröbner basis method [3].

Example 6. Consider the example Eq. (10)–(11). It can be verified that $(1, x, y, xy)^T$ is closed for multiplication with both x and y, meaning that the monomials x^2, y^2, $x^2 y$, and xy^2 can be written as linear functions of $(1, x, y, xy)^T$: From Eq. (10), we find $y^2 = x^2 - xy + 2y + 3$. After substitution of y^2 in Eq. (11), we find

$$x^2 = -1 + 3x + 3xy. \tag{35}$$

Multiplication of Eq. (35) with y yields

$$x^2 y - 3xy^2 = -y + 3xy. \tag{36}$$

From Eq. (11), we find $x^2 = 2y^2 - xy - 3x - 4y - 5$. After substitution of x^2 in Eq. (10), we find

$$y^2 = 2 + 2y + 2xy + 3x. \tag{37}$$

Multiplication of Eq. (37) with x yields

$$x^2y - 3xy^2 = -y + 3xy. \tag{38}$$

We can therefore write Eq. (35)–(38) as

$$\begin{pmatrix} 1 & 0 & 0 & 0 \\ 0 & 1 & 0 & 0 \\ -3 & 0 & -2 & 1 \\ 0 & 0 & 1 & -3 \end{pmatrix} \begin{pmatrix} x^2 \\ y^2 \\ x^2y \\ xy^2 \end{pmatrix} = \begin{pmatrix} -1 & 3 & 0 & 3 \\ 2 & 3 & 2 & 2 \\ 0 & 2 & 0 & 2 \\ 0 & 0 & -1 & 3 \end{pmatrix} \begin{pmatrix} 1 \\ x \\ y \\ xy \end{pmatrix}, \tag{39}$$

in which the 4×4 matrix in the left-hand side of the equation is invertible. This implies

$$\begin{pmatrix} x^2 \\ y^2 \\ x^2y \\ xy^2 \end{pmatrix} = \begin{pmatrix} -1 & 3 & 0 & 3 \\ 2 & 3 & 2 & 2 \\ 1.8 & -6.6 & 0.2 & -7.2 \\ 0.6 & -2.2 & 0.4 & -3.4 \end{pmatrix} \begin{pmatrix} 1 \\ x \\ y \\ xy \end{pmatrix}, \tag{40}$$

from which we easily find the equivalent eigenvalue problems

$$\mathbf{A}_x \mathbf{u} = x\,\mathbf{u}, \quad \text{and} \quad \mathbf{A}_y \mathbf{u} = y\,\mathbf{u}, \tag{41}$$

or

$$\begin{pmatrix} 0 & 1 & 0 & 0 \\ 0 & 0 & 0 & 1 \\ -1 & 3 & 0 & 3 \\ 1.8 & -6.6 & .2 & -7.2 \end{pmatrix} \begin{pmatrix} 1 \\ x \\ y \\ xy \end{pmatrix} = \begin{pmatrix} 1 \\ x \\ y \\ xy \end{pmatrix} x, \quad \text{and} \tag{42}$$

$$\begin{pmatrix} 0 & 0 & 1 & 0 \\ 2 & 3 & 2 & 2 \\ 0 & 0 & 0 & 1 \\ .6 & -2.2 & .4 & -3.4 \end{pmatrix} \begin{pmatrix} 1 \\ x \\ y \\ xy \end{pmatrix} = \begin{pmatrix} 1 \\ x \\ y \\ xy \end{pmatrix} y, \tag{43}$$

from which the solutions (x, y) follow, either from the eigenvalues or the eigenvectors. There are several other interesting properties we can deduce from this example. For example, since $\mathbf{A}_x$ and $\mathbf{A}_y$ share the same eigenvectors, they commute: $\mathbf{A}_x\mathbf{A}_y = \mathbf{A}_y\mathbf{A}_x$, and therefore also, any polynomial function of $\mathbf{A}_x$ and $\mathbf{A}_y$ will have the same eigenvectors.

2.6 Finding the Minimizing Root as a Maximal Eigenvalue

In many practical cases, and certainly in the polynomial optimization problem we started from in Section 2.2, we are interested in only one specific solution of the set of multivariate polynomials, namely the one that minimizes the polynomial objective function. As the problem is transformed into a (generalized) eigenvalue problem, we can now show that the minimal value of the given polynomial cost criterion corresponds to the maximal eigenvalue of the generalized eigenvalue problem.

Example 7. We illustrate this idea using a very simple optimization problem, shown in Fig. 2:

$$\min_{x,y} \quad J = x^2 + y^2 \tag{44}$$

$$\text{s. t.} \quad y = (x-1)^2, \tag{45}$$

the solution of which is found at $(x,y) = (0.41, 0.35)$ with a corresponding cost $J = 0.86$. The Lagrangian function is given by

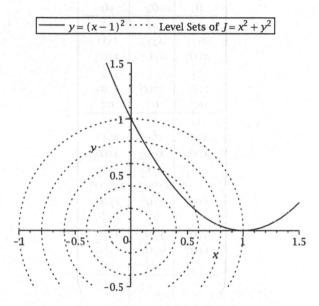

Fig. 2 A simple optimization problem, see Eq. (45). The constraint $y = (x-1)^2$ (solid line) and level sets of $J = x^2 + y^2$ (dotted lines) are shown. The solution is found at $(x,y) = (0.41, 0.35)$, for which $J = 0.86$.

$$\mathcal{L}(x,y,\alpha) = x^2 + y^2 + \alpha \left(y - (x-1)^2 \right). \tag{46}$$

The Lagrange multipliers method results in a system of polynomial equations in x, y, and α:

$$\frac{\partial \mathscr{L}}{\partial x}(x,y,\alpha) = 0 \qquad \Longleftrightarrow \qquad 2x - 2x\alpha + 2\alpha = 0, \qquad (47)$$

$$\frac{\partial \mathscr{L}}{\partial y}(x,y,\alpha) = 0 \qquad \Longleftrightarrow \qquad 2y + \alpha = 0, \qquad (48)$$

$$\frac{\partial \mathscr{L}}{\partial \alpha}(x,y,\alpha) = 0 \qquad \Longleftrightarrow \qquad y - x^2 + 2x - 1 = 0. \qquad (49)$$

Construct a matrix $\mathbf{M}_{d^*}$ as described above. In this example, $d^* = 4$, for which the dimension of $\mathbf{M}_4$ is 40×35. The truncated co-rank of $\mathbf{M}_{d^*}$ indicates there are three solutions. The corresponding 'canonical form' of the (truncated) kernel $\mathbf{V}_4$ is given by

$$\mathbf{V}_4 = \begin{pmatrix} 1 & 1 & 1 \\ x_1 & x_2 & x_3 \\ y_1 & y_2 & y_3 \\ \alpha_1 & \alpha_2 & \alpha_3 \\ x_1^2 & x_2^2 & x_3^2 \\ x_1 y_1 & x_2 y_2 & x_3 y_3 \\ x_1 \alpha_1 & x_2 \alpha_2 & x_3 \alpha_3 \\ y_1^2 & y_2^2 & y_3^2 \\ y_1 \alpha_1 & y_2 \alpha_2 & y_3 \alpha_3 \\ \alpha_1^2 & \alpha_2^2 & \alpha_3^2 \\ x_1^3 & x_2^3 & x_3^3 \\ x_1^2 y_1 & x_2^2 y_2 & x_3^2 y_3 \\ x_1^2 \alpha_1 & x_2^2 \alpha_2 & x_3^2 \alpha_3 \\ x_1 y_1^2 & x_2 y_2^2 & x_3 y_3^2 \\ x_1 y_1 \alpha_1 & x_2 y_2 \alpha_2 & x_3 y_3 \alpha_3 \\ x_1 \alpha_1^2 & x_2 \alpha_2^2 & x_3 \alpha_3^2 \\ y_1^3 & y_2^3 & y_3^3 \\ y_1^2 \alpha_1 & y_2^2 \alpha_2 & y_3^2 \alpha_3 \\ y_1 \alpha_1^2 & y_2 \alpha_2^2 & y_3 \alpha_3^2 \\ \alpha_1^3 & \alpha_2^3 & \alpha_3^3 \\ \vdots & \vdots & \vdots \end{pmatrix}, \qquad (50)$$

where (x_i, y_i, α_i) represent the roots (for $i = 1, \ldots, 3$). The objective function is given as $J = x^2 + y^2$. In accordance with the technique described in Section 2.5.1, we now define a diagonal matrix $\mathbf{D} = \mathrm{diag}(x_1^2 + y_1^2, \ x_2^2 + y_2^2, \ x_3^2 + y_3^2)$ containing the values of the objective function J evaluated at the roots. We have

$$\mathbf{S}_1 \mathbf{V}_4 \mathbf{D} = \mathbf{S}_2 \mathbf{V}_4, \quad \text{or} \quad \mathbf{S}_1 \mathbf{V}_4 = \mathbf{S}_2 \mathbf{V}_4 \mathbf{D}^{-1} \quad (\text{if} \quad J \neq 0), \qquad (51)$$

Again, $\mathbf{S}_1$ and $\mathbf{S}_2$ are row-selection matrices. In particular, $\mathbf{S}_1 \mathbf{V}_4$ selects the top rows of $\mathbf{V}_4$, whereas $\mathbf{S}_2 \mathbf{V}_4$ will make a linear combination of the suitable rows of $\mathbf{V}_4$ corresponding to the monomials of higher degrees in order to perform the multiplication with the objective function J. Again, the kernel of $\mathbf{M}_d$ is not directly available in the 'canonical form', instead, a kernel is computed as

$$\mathbf{M}_d \mathbf{Z}_d = 0, \tag{52}$$

where $\mathbf{Z}_d = \mathbf{V}_d \mathbf{T}$, for a certain non-singular matrix $\mathbf{T}$. Hence,

$$\mathbf{S}_1 \mathbf{Z}_d = \mathbf{S}_2 \mathbf{Z}_d (\mathbf{T}^{-1} \mathbf{D}^{-1} \mathbf{T}), \tag{53}$$

and the minimal norm is the inverse of the maximal eigenvalue of

$$(\mathbf{S}_2 \mathbf{Z}_d)^{\dagger} (\mathbf{S}_1 \mathbf{Z}_d) \quad (= \mathbf{T}^{-1} \mathbf{D}^{-1} \mathbf{T}), \tag{54}$$

where $\mathbf{X}^{\dagger}$ denotes the Moore-Penrose pseudoinverse of $\mathbf{X}$. This leads to the generalized eigenvalue problem

$$(\mathbf{Z}_d^T \mathbf{S}_2^T \mathbf{S}_1 \mathbf{Z}_d)\mathbf{u} = \lambda (\mathbf{Z}_d^T \mathbf{S}_2^T \mathbf{S}_2 \mathbf{Z}_d)\mathbf{u}, \tag{55}$$

where

$$\lambda = \frac{1}{J}. \tag{56}$$

By applying to $\mathbf{Z}_d$ the right similarity transformation (and a diagonal scaling), we also find $\mathbf{V}_d$.

We prefer to work with the power method for computing the maximum eigenvalue, instead of working with the inverse power method for the minimum eigenvalue, since in the inverse power method, a matrix inversion is required in each iteration step.

Example 8. We now apply this technique to example Eq. (10)–(11). We want to solve the following optimization problem (also see Fig. 1):

$$\min_{x,y} \quad J = x^2 + y^2 \tag{57}$$

$$\text{s. t.} \quad 3 + 2y - xy - y^2 + x^2 = 0 \tag{58}$$

$$5 + 4y + 3x + xy - 2y^2 + x^2 = 0. \tag{59}$$

The method of Lagrange multipliers now results in a set of four polynomial equations in four unknowns. The minimal degree $d^* = 5$ and corresponding matrix $\mathbf{M}_5$ of size 140×126 are found as described above. The cost criterion polynomial J is of degree two, which means that $\mathbf{S}_1 \mathbf{V}$ selects the parts of the basis of monomials corresponding to degrees 0 to $3(= 5 - 2)$. $\mathbf{S}_2$ is constructed so that the shift with $(x^2 + y^2)^{-1}$ is performed. The largest real eigenvalue from Eq. (55) yields the minimal cost as $1/0.1157 = 8.64$, which is attained at $(x, y) = (0.08, 2.93)$. This can be verified in Fig. 1, where the level sets of the cost function $J = x^2 + y^2$ are shown.

2.7 Algorithms

We have presented a technique to count the roots of systems of polynomial equations. Moreover, the roots can be determined from an eigenvalue problem. Since we search for the maximal eigenvalue of a certain matrix, a possible candidate algorithm is the power method [15].

Current methods for solving sets of polynomial equations can be categorized into symbolic, numerical, and hybrid types.

- One can classify symbolic techniques into Gröbner basis methods and resultant-based methods. The work on Gröbner bases [3] has dominated the field of algorithmic algebraic geometry for decades. In this approach, the original problem is transformed into an 'easier' equivalent, using symbolic techniques, such as Buchberger's algorithm [3]. However, Gröbner basis methods have fundamental disadvantages: they are restricted to small-scale problems, and moreover, the computations suffer from numerical instabilities, for example, two problems with seemingly small differences in coefficients can give rise to very differently looking Gröbner bases. Some efforts to bypass the costly generation of a Gröbner basis, by working towards a more direct formulation of a corresponding eigenvalue problem have been made, as in the use of border bases [33, 34]. On the other hand, resultant-based techniques are used to eliminate unknowns from a set of polynomial equations. Resultant-based methods are again gaining interest, as some of the disadvantages of Gröbner basis methods are solved, and the computations involved can be carried out using well-understood matrix computations.
- A wide variety of numerical solution techniques based on Newton's method have been developed. In general, methods based on Newton iterations fail to guarantee globally optimal solutions, but they can be used to find or refine a local solution, starting from an initial guess. Recently, in [20, 36, 38], relaxation methods for the global minimization of a polynomial, involving sums of squares and semidefinite programming have been presented. Many classical methods are often outperformed using this technique. However, in general, only a lower bound of the minimum of the objective function is found.
- Hybrid methods combine results from both the symbolic and the numerical perspectives to find all roots. Homotopy continuation methods [23, 43] track all solution paths, starting from an 'easy' problem through to the 'difficult' problem in question, hereby iterating prediction-correction steps based on Newton methods.

3 Applications in Systems Theory and Identification

The tasks of solving polynomial optimization problems and solving sets of polynomial equations are ubiquitous in science and engineering, and a wide variety of applications exists (e.g., computer vision [37], robotics: inverse kinematics [29], computational biology: conformation of molecules [13], life sciences: [2], etc.). The relation between systems theory and system identification, control theory, optimization theory, and algebraic geometry has only come to attention recently [4, 16, 18]. This perspective provides an ambitious and powerful framework to tackle many problems in the fields mentioned above. We highlight some important applications.

- The well-known prediction error methods (PEM) [24] can be phrased as optimization problems with a quadratic objective function and polynomial constraints representing the relations between the measured data and the model parameters. This means that they fit into the framework we have proposed above (cf. Section 2.2). The techniques presented here are quite ambitious, as they aim at finding global solutions to polynomial optimization problems. However, at this moment, the inherent complexity prohibits the application to large-scale problems.
- Least squares approximation of a matrix by a low-rank matrix is an important task in systems theory and identification, which can be solved using a singular value decomposition [12]. When additional constraints are imposed, e.g., linear matrix structure such as Hankel, or element-wise weighting, the so-called Riemannian singular value decomposition was proposed in [7, 8, 9, 10, 11] to solve the structured total least squares problem. The Riemannian SVD is essentially a system of polynomial equations, and can therefore be tackled using the methods described in this contribution. Moreover, the Riemannian SVD provides a unifying framework [22] for a multitude of existing and new system identification methods, e.g., prediction error methods: AR(X), ARMA(X), dynamical total least squares, errors-in-variables system identification, etc.
- In [17] some strong connections between polynomial system solving and multidimensional systems theory were revealed, especially between [40] and realization theory for multidimensional systems. In [16], many other interesting connections between constructive algebra and systems theory are established.
- In [26] it was shown that the question of assessing global identifiability for arbitrary (non-linear) model parametrizations is equivalent to the possibility of expressing the model structure as a linear regression: in [25], L. Ljung states:

> [The] result shows that the complex, non-convex form of the likelihood function with many local minima is not inherent in the model structure.

From the perspective of rearranging the identifiability question as a linear regression, the term 'algebraic convexification of system identification methods' was coined. Ljung and Glad use Ritt's algorithm, based on differential algebra, similar to the Gröbner basis algorithm [3]. Also this approach is related to solving systems of polynomial equations, and can be tackled using the techniques described in this paper.

4 Conclusions and Future Work

We have explored several fundamental links between systems theory and system identification, optimization theory, and algebraic geometry. We have generalized a technique based on Sylvester and Macaulay resultants, resulting in a method for root-counting as a rank test on the kernel of a Macaulay-like matrix. Also, the solutions can be determined: either as an eigenvalue problem, or by applying realization theory to the kernel of this matrix. In the case of a polynomial optimization problem, this technique can be applied to find the minimizing solution directly, by finding the maximal eigenvalue of a corresponding (generalized) eigenvalue problem.

The nature of this contribution is meant to be highly didactic, and presenting main ideas in a pedagogical way. We have omitted proofs and technical details, but yet, many practical challenges remain to be tackled before we can arrive at feasible numerical algorithms:

- How to go exactly from the rank deficiency of M_d to a (generalized) eigenvalue problem, needs to be investigated further. Moreover, in order to apply the power method, we need to prove that the largest eigenvalue, that is supposed to be equal to the inverse of the minimum of the objective function, is actually real; Said in other words, that there are no complex conjugated roots that in modulus are larger.
- Currently, many techniques similar to those described in this paper, suffer from a restrictive exponential complexity due to the rapidly growing number of monomials to be taken into account. This exponential complexity prohibits application to large problems. It remains to be investigated how the inherent complexity can be circumvented by exploiting the (quasi-Toeplitz) structure and sparsity.
- The relations between the technique presented here and the traditional symbolic methods will be investigated. The link with the work [14] is relevant in this respect.
- How results regarding rank tests, as observed in this article, are encountered in cases where the input polynomials describe a positive dimensional variety is also of interest.
- It is not completely clear how some properties observed in the Riemannian SVD framework can be exploited to devise efficient algorithms. For instance, all matrices encountered in this formulation are typically highly structured; It remains to be investigated how these properties might be exploited through the use of FFT-like techniques.

Acknowledgements Philippe Dreesen was supported by the Institute for the Promotion of Innovation through Science and Technology in Flanders (IWT-Vlaanderen). Research was also supported by Research Council KUL: GOA AMBioRICS, CoE EF/05/006 Optimization in Engineering (OPTEC), IOF-SCORES4CHEM, several PhD/postdoc & fellow grants; Flemish Government: FWO: PhD/postdoc grants, projects G.0452.04 (new quantum algorithms), G.0499.04 (Statistics), G.0211.05 (Nonlinear), G.0226.06 (cooperative systems and optimization), G.0321.06 (Tensors), G.0302.07 (SVM/Kernel), G.0320.08 (convex MPC), G.0558.08 (Robust MHE), G.0557.08 (Glycemia2), research communities (ICCoS, ANMMM, MLDM); IWT: PhD Grants, McKnow-E,

Eureka-Flite+; Helmholtz: viCERP; Belgian Federal Science Policy Office: IUAP P6/04 (DYSCO, Dynamical systems, control and optimization, 2007-2011); EU: ERNSI; Contract Research: AM-INAL.

References

[1] Auzinger, W., Stetter, H.J.: An elimination algorithm for the computation of all zeros of a system of multivariate polynomial equations. Proc. Int. Conf. Num. Math., pp. 11–30. Birkhäuser (1988)

[2] Barnett, M.P.: Computer algebra in the life sciences. ACM SIGSAM Bull. **36**, 5–32 (2002)

[3] Buchberger, B.: Ein algorithmus zum auffinden der basiselemente des restklassenringes nach einem nulldimensionalen polynomideal. Ph.D. thesis, University of Innsbruck (1965)

[4] Buchberger, B.: Gröbner bases and systems theory. Multidimens. Systems Signal Process. **12**, 223–251 (2001)

[5] Cox, D.A., Little, J.B., O'Shea, D.: Ideals, Varieties and Algorithms. Springer-Verlag (1997)

[6] Cox, D.A., Little, J.B., O'Shea, D.: Using Algebraic Geometry, second edn. Springer-Verlag, New York (2005)

[7] De Moor, B.: Structured total least squares and L_2 approximation problems. Lin. Alg. Appl. **188/189**, 163–207 (1993)

[8] De Moor, B.: Dynamic total linear least squares. In: Proc. 10th IFAC Symp. System Identif., vol 3, pp. 159–164. Copenhagen, Denmark (1994)

[9] De Moor, B.: Total least squares for affinely structured matrices and the noisy realization problem. IEEE Trans. Signal Process. **42**(11), 3104–3113 (1994)

[10] De Moor, B.: Linear system identification, structured total least squares and the Riemannian SVD. In: S. Van Huffel (ed.) Recent advances in Total Least Squares Techniques and Errors-In-Variables modeling, pp. 225–238. SIAM, Philadelphia (1997)

[11] De Moor, B., Roorda, B.: L_2-optimal linear system identification: Structured total least squares for SISO systems. Proc. 33rd Conf. Decis. Control (CDC), Lake Buena Vista, FL (1994)

[12] Eckart, G., Young, G.: The approximation of one matrix by another of lower rank. Psychometrika **1**, 211–218 (1936)

[13] Emiris, I.Z., Mourrain, B.: Computer algebra methods for studying and computing molecular conformations. Algorithmica **25**, 372–402 (1999)

[14] Faugère, J.C.: A new efficient algorithm for computing Gröbner bases (F4). J. Pure Appl. Algebra **139**(1), 61–88 (1999)

[15] Golub, G.H., Van Loan, C.F.: Matrix Computations, third edn. Johns Hopkins University Press, Baltimore, MD, USA (1996)

[16] Hanzon, B., Hazewinkel, M. (eds.): Constructive Algebra and Systems Theory. Royal Netherlands Academy of Arts and Sciences (2006)

[17] Hanzon, B., Hazewinkel, M.: An introduction to constructive algebra and systems theory. In: B. Hanzon, M. Hazewinkel (eds.) Constructive Algebra and Systems Theory, pp. 2–7. Royal Netherlands Academy of Arts and Sciences (2006)

[18] Hanzon, B., Jibetean, D.: Global minimization of a multivariate polynomial using matrix methods. J. Glob. Optim. **27**, 1–23 (2003)

[19] Jónsson, G.F., Vavasis, S.A.: Accurate solution of polynomial equations using Macaulay resultant matrices. Math. Comput. **74**(249), 221–262 (2004)

[20] Lasserre, J.B.: Global optimization with polynomials and the problem of moments. SIAM J. Optim **11**(3), 796–817 (2001)

[21] Lazard, D.: Résolution des systèmes d'équations algébriques. Theor. Comput. Sci. **15**, 77–110 (1981)

[22] Lemmerling, P., De Moor, B.: Misfit versus latency. Automatica **37**, 2057–2067 (2001)

[23] Li, T.Y.: Numerical solution of multivariate polynomial systems by homotopy continuation methods. Acta Numer. **6**, 399–436 (1997)

[24] Ljung, L.: System identification: theory for the user. Prentice Hall PTR, Upper Saddle River, NJ (1999)

[25] Ljung, L.: Perspectives on system identification. Proc. 17th IFAC World Congress, pp. 7172–7184. Seoul, Korea (2008)

[26] Ljung, L., Glad, T.: On global identifiability for arbitrary model parametrizations. Automatica **35**(2), 265–276 (1994)

[27] Macaulay, F.S.: On some formulae in elimination. Proc. London Math. Soc. **35**, 3–27 (1902)

[28] Macaulay, F.S.: The algebraic theory of modular systems. Cambridge University Press (1916)

[29] Manocha, D.: Algebraic and numeric techniques for modeling and robotics. Ph.D. thesis, Computer Science Division, Department of Electrical Engineering and Computer Science, University of California, Berkeley (1992)

[30] Manocha, D.: Solving systems of polynomial equations. IEEE Comput. Graph. Appl. **14**(2), 46–55 (1994)

[31] Möller, H.M., Stetter, H.J.: Multivariate polynomial equations with multiple zeros solved by matrix eigenproblems. Numer. Math. **70**, 311–329 (1995)

[32] Mourrain, B.: Computing the isolated roots by matrix methods. J. Symb. Comput. **26**, 715–738 (1998)

[33] Mourrain, B.: A new criterion for normal form algorithms. In: Applied Algebra, Algebraic Algorithms and Error-Correcting Codes, *Lecture Notes in Computer Science*, vol. 1710, pp. 430–443. Springer, Berlin (1999)

[34] Mourrain, B., Trébuchet, P.: Solving complete intersections faster. Proc. ISSAC 2000, pp. 231–238. ACM, New York (2000)

[35] Nocedal, J., Wright, S.J.: Numerical Optimization, second edn. Springer (2006)

[36] Parrilo, P.A.: Structured semidefinite programs and semialgebraic geometry methods in robustness and optimization. Ph.D. thesis, California Institute of Technology (2000)

[37] Petitjean, S.: Algebraic geometry and computer vision: Polynomial systems, real and complex roots. J. Math. Imaging Vis. **10**, 191–220 (1999)

[38] Shor, N.Z.: Class of global minimum bounds of polynomial functions. Cybernetics **23**(6), 731–734 (1987)

[39] Stetter, H.J.: Matrix eigenproblems are at the heart of polynomial system solving. ACM SIGSAM Bull. **30**(4), 22–25 (1996)

[40] Stetter, H.J.: Numerical Polynomial Algebra. SIAM (2004)

[41] Stetter, H.J.: An introduction to the numerical analysis of multivariate polynomial systems. In: B. Hanzon, M. Hazewinkel (eds.) Constructive Algebra and Systems Theory, pp. 35–47. Royal Netherlands Academy of Arts and Sciences (2006)

[42] Sylvester, J.J.: On a theory of syzygetic relations of two rational integral functions, comprising an application to the theory of sturms function and that of the greatest algebraical common measure. Trans. Roy. Soc. Lond. (1853)

[43] Verschelde, J., Verlinden, J., Cools, R.: Homotopies exploiting Newton polytopes for solving sparse polynomial systems. SIAM J. Numer. Anal. **31**, 915–930 (1994)

Part II
Bridging Theory and Applied Technology

Designing Instrumentation for Control

Faming Li, Maurício C. de Oliveira and Robert E. Skelton

Abstract For linear time-invariant systems with full order feedback controllers, this paper completely solves the problem of selecting sensor and actuator precisions required to yield specified upper bounds on control covariances, output covariances, and precision constraints. Moreover, the paper proves that this problem is convex. The contribution here is that selection of sensors and actuators can now be a part of the control design problem, in lieu of the classical control problem, where sensor and actuator resources have already been specified before the "control" problem is defined. Furthermore, an *ad hoc* algorithm is given to select which sensors and actuators should be selected for the final control design that constrains control covariance, output covariance, and a financial cost constraint (where we assume that precision is linearly related to financial cost). We label this the Economic Design Problem.

1 Motivation

The effectiveness of any control system critically depends upon the information provided about the *dynamics* (from which model we predict the motion) and the information provided about the *actual motion* (which we obtain from sensors). The information about dynamics is usually provided in the form of a set of ordinary differential equations that must be determined as an *appropriate* approximation of the actual motion, and the information provided about the actual motion comes from an *appropriate* selection of sensors. Neither of these two critical issues in control theory (to find *appropriate* models and *appropriate* real-time information) have been

Faming Li
Xerox Research Center, Webster, NY USA e-mail: Faming.Li@XEROX.com

Maurício C. de Oliveira and Robert E. Skelton
Department of Mechanical and Aerospace Engineering, University of California San Diego,
La Jolla, CA 92093-0411, USA e-mail: {mauricio,bobskelton}@ucsd.edu

P.M.J. Van den Hof et al. (eds.), *Model-Based Control: Bridging Rigorous Theory and Advanced Technology*, DOI: 10.1007/978-1-4419-0895-7_5,
© Springer Science + Business Media, LLC 2009

properly resolved. Theory is absent to help select the most *appropriate* model to use for the control design, and theory is absent to select the sensors and the actuators to achieve the best performance. More importantly theory is absent to guide one to design the plant in the first place, to make it more easily controlled to the required performance, with the least control effort.

Such are the goals of a true *systems design theory*, which is presently not available, and probably a complete design theory (for jointly selecting all the system components) will never be available. Nonetheless, important milestones can be attained in these directions. The vast majority of the control literature available today involves design problems conveniently defined to search for a controller *after* the sensors and actuators have been selected, *after* the system model has been obtained, and most certainly *after* the plant (to be controlled) has been designed. Unfortunately, these are not independent problems, even though university classes might give students the opposite impression.

Universities teach component technologies, drawing narrow boundaries around subjects for tractability, or easy management of research groups, leading to uncoordinated decisions about system design. We separate the decisions about manufacturing, modeling, information architecture, and control. We have not yet learned how to take Michael Faraday's advice "We should begin with the whole, then construct the parts." By *system design* we refer to the task of determining the design requirements of the components (plant components, sensors, actuators, controller, given only the requirements of the final system.

The first requirement for a comprehensive system design strategy is to find a procedure by which one can determine the performance-limiting technology in our system. Perhaps we could get better performance by improving the manufacturing precision of components. If so, which components? Perhaps we could get better performance by improving the model of the system components. If so, which components? Better performance might result from improved precision or distribution of sensing or actuating. If so, which instruments? Maybe we are constrained due to computational precision, or maybe by the algorithm being used for signal processing. More money and resources spent on any of these components might be wasted if we are already operating at the fundamental limits of physics. A systematic process to answer any of these questions is not available. We have no procedure to decide where to spend money effectively, to improve system design. There is a common fallacy that underlies many funding agencies, that "better systems are obtained by better components" (and so funding agencies should only develop procedures to improve components). Unfortunately improved precision of a component may not improve the system, or may do so at an unreasonable cost in money or energy.

Consider the specific problems concerned with limits on computational capability and control design. Others have described the intersection of signal processing and control design for a given controller, but Okko Bosgra was the first to make a substantial contribution to the more difficult challenge related to a fixed order controller driving a full-order plant. His paper was the pioneering work of output feedback control theory, even though many authors failed to reference Bosgra's pioneering paper [12]. This paper was a powerful motivator for me to look at integrating modeling and control design and related topics over many years. I benefited greatly from his insights.

For a linear time-invariant plant model, this paper seeks to integrate the selection of control gains and the selection of sensor/actuator precision. This is a small step toward a system design theory as discussed above.

2 Definition of Information Architecture

For the purpose of this discussion let us define the phrase *control architecture* as the structure of the controller (perhaps decentralized in a specific manner), dictating which group of sensors communicate with which group of actuators. Likewise we will label the following set of decisions as fixing the *information architecture*:

Instrument type: Choose the instrument (actuator or sensor) functionality (to exert or measure force, velocity, etc).

Instrument precision: Choose the signal-to-noise ratio of the instruments and/or of the software processing the instrument measurements.

Instrument location: Choose the variables in the plant that should be manipulated or measured. For example, in the control or estimation of a mechanical structure, this is equivalent to choosing the *location* of the actuators and sensors.

Of course, these decisions cannot be made sensibly without also selecting the control or estimation algorithms. Hence, to the above list we *include* the design of the control or estimation algorithm as a necessary part of selecting the *information architecture*. Selecting the *information architecture* is the subject of this paper.

We will address two problems in this paper. First we will solve the *information architecture* problem in the case where the location of all sensors and actuators are *fixed*. This first problem is completely solved and shown to be convex. We choose the precision of the sensors/actuators such that closed loop performance (covariance upper bound) is guaranteed. Actually we choose the *worst* precision (cheapest solution) that allows acceptable closed loop performance. Since we presume that cost is proportional to precision, we label this technique *economic system design*. Secondly we address the complete information architecture problem, where we will choose an algorithm to select a reduced set of sensors and actuators. There is an *ad hoc* step in this second problem.

3 Background

A number of papers dealt with the optimal sensor/actuator placement problem in the framework of integrated control-structure design [3, 7, 13, 17]. Perhaps most papers on the subject on sensor/actuator location have focused on optimizing degrees of observability or controllability, for example Gawronski and Lim [3]. It is important to state at the beginning of this paper that locating sensors and actuators in a control problem is not a study about observability and controllability. It is about *performance* (remembering Michael Faraday's advice). Yes, one can optimize controllability in step 1 and observability in step 2, but this strategy ignores the fact that increasing observability of dynamics that do not need to be controlled is not helpful, and in fact counterproductive. Likewise increasing controllability of dynamics that contribute little to the required control performance is not helpful. Performance (say bounds on output covariance) is more closely related to *products* of observability and observability measures, rather than individual controllability and observability measures alone. But other scaling factors enter the scene when trying to decompose the contribution of a sensor or actuator in the overall performance measure, hence controllability and observability measures are not especially useful here.

In Skelton and DeLorenzo[17] and Skelton and Li [19], a LQG (Linear Quadratic Gaussian optimal control) type cost function is decomposed into contributions of each sensor and each sensor/actuator. Sensors/actuators are selected for removal if the contribution in the cost function is small. The disadvantage of this method is that this is a cost decomposition strategy rather than a cost composition strategy. All the component costs change after deletion of any component. No information is provided that suggests how the total performance will be affected by the deletion of a component. Chmielewski, Palmer, and Manousiouthakis [2] proposed a procedure to select only sensors based on estimation error covariance. Although economy has always been a design concern, a tractable way of including economics in control design appeared only recently, in Lu and Skelton [10], by assigning a price to sensors and actuators proportional to their precision (signal-to-noise ratio). In the so-called *Economic Design Problem*, the metric to be minimized is defined as the total instrument cost (where the cost of each instrument is assumed to be proportional to the signal-to-noise ratio of each instrument), subject to covariance upper bounds on outputs. The proposed schemes are iterative, and no optimality can be guaranteed.

4 Contributions of this Paper

Past research has not provided a globally optimal solution to simultaneously select controls and sensor/actuator hardware precision. This paper remedies that situation. The construction of the paper is as follows.

First, a system design problem is set up as a trade-off among three competitive factors: control performance, control effort and sensor/actuator precision. We solve this problem completely, with a guaranteed global optimum obtained through convex optimization (more precisely an LMI – Linear Matrix Inequality) in the case of state feedback or full order dynamic output feedback controllers. Note that in this step we do not *select* sensors and actuators. Choosing the required precision for sensors and actuators simultaneously with controller selection is important, since these are obviously interdependent problems. So the good news is that increasing the domain of the optimization allows the problem to remain convex even if one wishes to design the sensor and actuator precision as well. This problem is a particular case of the multi-objective robust control addressed in [15, 18] which, in general, does not reduce to a convex problem.

The second task is to decide which sensors and which actuators to delete from the larger set. This step is *ad hoc* and the proposed algorithm is described in simple terms as follows:

Step 1. Insert the entire set of feasible sensors and actuators in the system model, but allow the precision of each device be free to choose.
Step 2. Solve a convex problem to choose the control gains and the sensor/actuator precision. This produces the controller for this set of sensors/actuators with some model-based controller design method.
Step 3. Delete the sensor or actuator requiring the smallest precision (largest noise) if there is a large gap between the precisions required. Otherwise, stop.

The sensor/actuator selection method proposed in the present paper generalizes the previous work [19], which uses the maximal accuracy as a performance constraint.

5 Problem Statement

Let a continuous-time time invariant linear system be described in state-space by the following equations

$$\dot{x} = Ax + Bu + D_p w_p + D_a w_a \quad \text{(Plant)}$$
$$y = Cx \quad \text{(Output)} \tag{1}$$
$$z = Mx + E w_s \quad \text{(Measurement)}$$

In the above, x is the state of the system to be controlled, u is a vector of control inputs and z is a vector of measurements. The vector y is the output of the system we wish to regulate and the w's are noisy input vectors. We assume that E is full rank.

In the above model, the different w's will be used to model actuator noise (w_a), sensor noise (w_s) and ambient noise (w_p). In this paper w_a, w_s and w_p are modeled as independent zero mean Gaussian white noises with covariances W_a, W_s and W_p, respectively. That is,

$$\mathcal{E}_\infty(w_i) = 0, \qquad \mathcal{E}_\infty(w_i w_i^T) = W_i \delta(t - \tau), \qquad i = \{a, s, p\},$$

where $\mathcal{E}_\infty(x) = \lim_{t \to \infty} \mathcal{E}(x)$ denotes the asymptotic expected value of the random variable x.

In the control literature it is standard to assume that the covariance matrices W_i, $i = \{a, s, p\}$ are known. Perhaps the main innovation in the present work is to let W_i, $i = \{a, s\}$, that is the covariances associated with sensor and actuator noise to be unknown, and let the optimization problem inform the user of a required value of accuracy associated with these variances.

More specifically, we assume W_p to be known and fixed and

$$\Gamma_a \triangleq W_a^{-1}, \qquad\qquad \Gamma_s \triangleq W_s^{-1}$$

to be determined by an optimization problem. For simplicity and other practical considerations that will be explained later, we assume that W_a and W_s are diagonal matrices (hence Γ_a and Γ_s are also diagonal), i.e. that sensor and actuator noise that enter the system are not correlated between different sensors and actuators. Of course they become correlated as they are processed by the system. For convenience, we also define vectors γ_a and γ_s such that

$$\Gamma_a = \mathrm{diag}(\gamma_a), \qquad\qquad \Gamma_s = \mathrm{diag}(\gamma_s)$$

as γ_a and γ_s will be used as decision variables.

The main contribution of this paper is to let the precision of each sensor/actuator (i.e. the inverse of their noise variances), which have been grouped in γ_a and γ_p, to be determined along with the control input. Of course, it is natural to expect that for a fixed set of sensors and actuators the best possible control will be achieved with the maximum possible instrument precision. This is certainly the case in the framework of this paper. For such reason, we associate to each instrument a "price", which we assume to be a linear function of the γ's, that is, inversely proportional to the instrument noise intensities. We shall label the resulting overall design procedure an *economic design*, as per [9, 10]. A total instrument price can be expressed as

$$\$ = p_a^T \gamma_a + p_s^T \gamma_s$$

where p_a and p_s are vectors containing the price of each instrument per unit of precision. Hence '$\$$' is the total economic cost of all the sensors and actuators.

As for control design, we shall seek to determine a linear dynamic output-feedback controller of the form

$$\begin{aligned}
\dot{x}_c &= A_c x_c + B_c z \\
u &= C_c x_c + D_c z
\end{aligned} \tag{2}$$

The problem to be solved is defined as follows.

Integrated sensor/actuator selection and control design problem
Design a dynamic output feedback controller and allocate sensor/actuator precisions such that the following constraints are satisfied:

$$\begin{aligned}
\$ < \bar{\$}, \quad \gamma_a < \bar{\gamma}_a, \quad \gamma_s < \bar{\gamma}_s, \\
\mathcal{E}_\infty\left(uu^T\right) < \bar{U}, \quad \mathcal{E}_\infty\left(yy^T\right) < \bar{Y},
\end{aligned} \tag{3}$$

for given specifications $\bar{\$}$, $\bar{\gamma}_a$, $\bar{\gamma}_s$, $\bar{U}$ and $\bar{Y}$.

These system performance constraints are represented by bounds on the control and measured output variances ($\bar{U}$ and $\bar{Y}$). The bounds $\bar{\gamma}_a$ and $\bar{\gamma}_s$ represent limits of precision on each available instrument, as provided by the manufacturer. Each entry of $\bar{\gamma}_a$ and $\bar{\gamma}_s$ indicates the best precision available in the marketplace. Finally, as in practice, the instrument budget might be constrained, a budget bound $\bar{\$}$ is present.

Later in this paper we shall also consider a variant of the above problem we call the **economic design problem**. The objective in this problem is that of minimizing the total instrument price \$ while satisfying the other system performance constraints.

6 Solution to the General Integrated Sensor/Actuator Selection and Control Design Problem

In this section we provide a solution to the control problem stated in the previous section in the form of Linear Matrix Inequalities (LMI). The set of feasible solutions of an LMI is convex and solutions to such feasibility problems can be found in polynomial time [1].

One might expect that the problem of designing a controller subject to one of the individual constraints in (3) can be converted into a convex problem using standard methods in the literature, e.g. [18]. In the presence of multiple constraints, control design problems are often not convex. Under certain assumptions, convex suboptimal solutions have been proposed, e.g. [5, 14, 15]. Here we will show that a solution to the integrated sensor/actuator selection and control design, i.e. in the presence of all constraints (3), can still be computed by solving a certain convex LMI problem. This LMI problem is presented in the next theorem. For convenience of notation we use the symbol $(\cdot)^T$ on an off-diagonal block indicates the transpose of a corresponding symmetric block and $\text{Sym}(X)$ as short for '$X + X^T$'.

Theorem 1. *Let a continuous-time time-invariant linear system be described by the state space equations (1). There exists controller matrices (A_c, B_c, C_c, D_c) such that the cost and performance constraints (3) are satisfied if and only if there exist symmetric matrices X, Y, vectors γ_a, γ_s, and matrices L, F, and Q such that*

$$p_a^T \gamma_a + p_s^T \gamma_s < \bar{\$}, \qquad \gamma_a < \bar{\gamma}_a, \qquad \gamma_s < \bar{\gamma}_s, \tag{4}$$

$$\begin{bmatrix} \bar{Y} & CX & C \\ (\cdot)^T & X & I \\ (\cdot)^T & (\cdot)^T & Y \end{bmatrix} > 0, \qquad \begin{bmatrix} \bar{U} & L & 0 \\ (\cdot)^T & X & I \\ (\cdot)^T & (\cdot)^T & Y \end{bmatrix} > 0, \qquad \begin{bmatrix} \text{Sym}(\Phi_{11}) & \Phi_{12} \\ (\cdot)^T & \Phi_{22} \end{bmatrix} < 0, \tag{5}$$

where

$$\Phi_{11} := \begin{bmatrix} AX + BL & A \\ Q & YA + FM \end{bmatrix}, \qquad \Phi_{21} := \begin{bmatrix} D_p & D_a & 0 \\ YD_p & YD_a & FE \end{bmatrix},$$

and $\Phi_{22} := -\operatorname{diag}(W_p^{-1}, \Gamma_a, \Gamma_s)$. *If the above LMI has a feasible solution then suitable controller matrices can be computed from the solution as*

$$\begin{bmatrix} A_c & B_c \\ C_c & D_c \end{bmatrix} = \begin{bmatrix} V^{-1} & -V^{-1}YB \\ 0 & I \end{bmatrix} \times \begin{bmatrix} Q - YAX & F \\ L & 0 \end{bmatrix} \begin{bmatrix} U^{-1} & 0 \\ -MXU^{-1} & I \end{bmatrix}, \qquad (6)$$

where V and U are nonsingular square matrices satisfying $YX + VU = I$.

Proof. See [6].

7 Particular Cases of the Integrated Sensor/Actuator Selection and Control Design Problem

7.1 State feedback control

In case that a noiseless measurement of the state x is available for feedback, as opposed to the noisy measurement vector z, the optimal controller that solves the Integrated Sensor/Actuator Selection and Control Design Problem becomes a static controller of the form $u = Kx$. The solution to this problem is obtained by solving a simplified version of the LMI problem defined in Theorem 1, given in the next corollary.

Corollary 1. *Let a continuous-time time-invariant linear system be described by the state space equations (1) in which a noiseless measurement of the state vector is available, i.e. $z = x$. There exists a state feedback controller $u = Kx$ such that the constraints*

$$\$ < \bar{\$}, \quad \gamma_a < \bar{\gamma}_a, \quad \mathcal{E}_\infty\left(uu^T\right) < \bar{U}, \quad \mathcal{E}_\infty\left(yy^T\right) < \bar{Y}, \qquad (7)$$

for given $\bar{\$}$, $\bar{\gamma}_a$, $\bar{U}$ and $\bar{Y}$, are satisfied if and only if there exist a symmetric matrix X, a vector γ_a and a matrix L such that

$$p_a^T \gamma_a < \bar{\$}, \qquad \gamma_a < \bar{\gamma}_a, \qquad CXC^T < \bar{Y}, \qquad (8)$$

$$\begin{bmatrix} \bar{U} & L \\ (\cdot)^T & X \end{bmatrix} > 0, \qquad \begin{bmatrix} \operatorname{Sym}(AX + BL) & D_p & D_a \\ (\cdot)^T & -W_a^{-1} & 0 \\ (\cdot)^T & 0 & -\Gamma_a \end{bmatrix} < 0. \qquad (9)$$

If the above LMI has a feasible solution then a suitable state feedback controller gain can be computed as $K = LX^{-1}$.

Proof. Follows from Theorem 1 using standard arguments, e.g. [4].

The importance of this problem in the context of this paper is that it provides the limits of performance achievable with perfect sensing. In other words, it provides the optimal control law and actuator selection in the presence of perfect measurements. This is valuable information that can help identify actuation requirements. Note also that in the event that a noisy state vector is available, i.e. $z = x + Ew_s$, then the optimal controller is not static but dynamic and should be computed from Theorem 1.

7.2 Estimation

The dual version of the state feedback control problem stated in the previous section is the optimal state estimation. The estimation problem considered here is that of determining a gain G and the *state observer*

$$\dot{\hat{x}} = A\hat{x} + Bu + G(z - \hat{z}),$$
$$\hat{z} = M\hat{x}, \tag{10}$$
$$e = C(x - \hat{x})$$

such that all constraints in (3) are satisfied, where the constraint $\mathscr{E}_\infty(yy^T) < \bar{Y}$ is replaced by a constraint on the covariance of the output estimation error e, namely $\mathscr{E}_\infty(ee^T) < \bar{Y}$. In the context of state estimation, the control input u is supposed to be precisely known, so that the constraint on the covariance of u can be taken as given. A solution to this problem is given in the next corollary to Theorem 1.

Corollary 2. *Let a continuous-time time-invariant linear system be described by the state space equations (1). There exists a state observer (10) and a gain G such that the constraints*

$$\$ < \bar{\$}, \qquad \gamma_a < \bar{\gamma}_a, \qquad \gamma_s < \bar{\gamma}_s, \qquad \mathscr{E}_\infty(ee^T) < \bar{Y}, \tag{11}$$

for given $\bar{\$}$, $\bar{\gamma}_a$, $\bar{\gamma}_s$ and $\bar{Y}$, are satisfied if and only if there exist a symmetric matrix Y, vectors γ_a, γ_s, and a matrix F such that

$$p_a^T \gamma_a + p_s^T \gamma_s < \bar{\$}, \qquad \gamma_a < \bar{\gamma}_a, \qquad \gamma_s < \bar{\gamma}_s, \tag{12}$$

$$\begin{bmatrix} \bar{Y} & C \\ (\cdot)^T & Y \end{bmatrix} > 0, \qquad \begin{bmatrix} \mathrm{Sym}(YA + FM) & YD_p & YD_a & FE \\ (\cdot)^T & -W_p^{-1} & 0 & 0 \\ (\cdot)^T & 0 & -\Gamma_a & 0 \\ (\cdot)^T & 0 & 0 & -\Gamma_s \end{bmatrix} < 0. \tag{13}$$

If the above LMI has a feasible solution then a suitable observer gain can be computed as $G = Y^{-1}F$.

Proof. Follows from Theorem 1 using standard arguments, e.g. [4].

As for state feedback control, the importance of the above problem is in establishing limits on the best achievable control performance. In output feedback control design, the actuator and measurement noises will deteriorate the output performance no matter how large is the control effort. For instance, in a regulator problem, there exists steady state output error due to the actuator and measurement noises. Maximal accuracy characterizes the best performance one can attain with the existing actuator and measurement noises. In the above problem, if one lets $\bar{Y}$ be a variable and minimizes $\text{trace}(\bar{Y})$, the solution will result in the best actuator and sensor precision for maximum accuracy given cost and precision constraints. Note that contrary to what happens with state feedback control, in our setting, the actuator noise plays a role in determining maximum accuracy. This is a consequence of the fact that the actuator noise enters the plant directly, whereas sensor noise is filtered by the controller before entering the plant.

Hence, maximal accuracy reflects the precision of both the actuators and sensors. When the actuators are specified, that is, $\hat{W}_a$ is given, the maximal accuracy reduces to estimation error and depends on the sensor precisions only. Finally, note that maximal accuracy is a measure of the best possible output performance regardless of control effort. Hence it can be a conservative performance indicator when the system allows big output error.

7.3 Economic design problem

In this section we revisit the economic design problem [9, 10]. That is, we focus on the problem of minimizing the total cost \$ while satisfying the system performance constraints in (3). In the light of Theorem 1, a solution to the economic design problem is given by solving the convex problem

$$\min_{\bar{\$},X,Y,\gamma_a,\gamma_s,L,F,Q} \{\bar{\$}: \quad (4),(5)\}. \tag{14}$$

This is in contrast with the more involved iterative procedure proposed in [9, 10]. The key to the simplification is the reformulation of the constraints as LMI instead of the Riccati equations used in [9, 10]. In Section 9 we will use the economic design problem as a means to select sensors and actuators.

8 Discrete-time systems

Until this point we have considered the integrated sensor/actuator selection and control design problem for continuous-time systems. The version of this problem for discrete-time systems has some subtle differences that deserve special consideration.

First, the continuous-time closed-loop system must be strictly proper, in order for the variances of interest, i.e. the output and control input variances, to be finite. This is best achieved by constraining the controller to be strictly proper as well.

For discrete-time systems, the asymptotic variances are finite even for non-strictly proper systems, and a non-strictly proper controller may provide some extra degrees of freedom in the design. This implies small modifications on the LMIs as compared with the continuous-time Theorem 1 which will be addressed in Theorem 2.

Second, and perhaps more important, in controlling continuous-time systems with a discrete-time controller, noise can be added to the closed-loop system at the digital-to-analog converter (DAC) and the analog-to-digital converter (ADC) in the form of quantization errors. Several works in the literature support modeling such quantization noise as white noise, e.g. [8, 11, 20, 21]. Under this assumption, the same idea used in the case of pure sensor and actuator noise can be used to model quantization error at the digital/analog interfaces, the resulting integrated sensor/actuator selection and control design problem being simply augmented by noise sources associated with the DAC/ADC converters. This provides an avenue to *allocate computational resources,* by letting the control optimization select the adequate number of bits required to achieve a specified system performance as well as the sensors and actuators. A reduction of the number of bits per channel is also likely to reduce the needs for communication in case the control systems is done through a computer network, a problem much in vogue these days. The modified LMIs for discrete-time systems in given in the next theorem.

Theorem 2. *Let a discrete-time time-invariant linear system be described by the state space equations (1) in which the derivative is replaces by a forward time-shift. There exists controller matrices (A_c, B_c, C_c, D_c) such that the cost and performance constraints (3) are satisfied if and only if there exist symmetric matrices X, Y, vectors γ_a, γ_s, and matrices L, F, Q, R such that*

$$p_a^T \gamma_a + p_s^T \gamma_s < \bar{\$}, \qquad \gamma_a < \bar{\gamma}_a, \qquad \gamma_s < \bar{\gamma}_s, \tag{15}$$

$$\begin{bmatrix} \bar{Y} & CX & C \\ (\cdot)^T & X & I \\ (\cdot)^T & (\cdot)^T & Y \end{bmatrix} > 0, \quad \begin{bmatrix} \bar{U} & L & RM & RE \\ (\cdot)^T & X & I & 0 \\ (\cdot)^T & (\cdot)^T & Y & 0 \\ (\cdot)^T & (\cdot)^T & (\cdot)^T & \Gamma_s \end{bmatrix} > 0, \quad \begin{bmatrix} \Psi_{11} & \Psi_{12} & \Psi_{13} \\ (\cdot)^T & \Psi_{11} & 0 \\ (\cdot)^T & (\cdot)^T & \Psi_{33} \end{bmatrix} > 0 \tag{16}$$

where

$$\Psi_{11} := \begin{bmatrix} X & I \\ I & Y \end{bmatrix}, \quad \Psi_{12} := \begin{bmatrix} AX + BL & A + BRM \\ Q & YA + FM \end{bmatrix}, \quad \Psi_{13} := \begin{bmatrix} D_p & D_a & BRE \\ YD_p & YD_a & FE \end{bmatrix},$$

and $\Psi_{33} := \Phi_{22}$, as defined in Theorem 1. If the above LMI has a feasible solution then suitable controller matrices can be computed from the solution as

$$\begin{bmatrix} A_c & B_c \\ C_c & D_c \end{bmatrix} = \begin{bmatrix} V^{-1} & -V^{-1}YB \\ 0 & I \end{bmatrix} \times \begin{bmatrix} Q - YAX & F \\ L & R \end{bmatrix} \begin{bmatrix} U^{-1} & 0 \\ -MXU^{-1} & I \end{bmatrix}, \tag{17}$$

where V and U are nonsingular square matrices satisfying $YX + VU = I$.

Proof. See [6]

Note that in the discrete-time the controller may not be strictly proper. If a strictly proper controller is desired one can add the linear constraint $R = 0$ to the above inequalities without destroying the problem convexity.

9 Sensor and Actuator Selection

One has often to decide on the *precision* required by each sensor and actuator but also on the *number* and *location* of sensors and actuators. In this section, a heuristic design procedure is proposed for the selection of actuators and sensors: we start with all possible actuators and sensors, then solve the integrated sensor/actuator selection and control design problem to identify the sensor or actuator that requires the least precision. By deleting that sensor/actuator each time till loss of feasibility, one is left with a system with a smaller number of sensors and actuators whose precision has been selected to match a given performance constraint.

The problem we solve at each iteration is the economic design problem, as defined in (14). We simultaneously select the instruments and design a controller in order to minimize the instrument cost. As seen in Section 7.3, Theorem 1 solves the economic design problem and produces the instrument precision vectors γ_a and γ_s. At each iterations we shall sort the elements of the resulting vector (γ_a, γ_s) in descending order, so that the last instrument requires the least precision. We now know where to spend money, making the components corresponding to large entries reliable, because performance is critical to these components. The components corresponding to the least entries might be taken off the shelf, or even deleted. In our algorithm, we shall delete the least demanding sensor/actuator each time and repeat the minimization till loss of feasibility. And for each selection of sensor/actuator, the economic design is carried out. In conclusion, we state the economic instrument selection algorithm as follows:

Algorithm 1: The Economic Instrument Selection Algorithm.

Step 1.   Place all admissible types of sensors/actuators at all admissible locations. Select a cost upper bound ($\bar{\$}$), an output covariance upper bound ($\bar{Y}$), a control covariance upper bound ($\bar{U}$), and the highest available instrument precisions ($\bar{\gamma}_a$ and $\bar{\gamma}_s$).

Step 2.   Solve the Economic Design Problem (14). This yields the controller (A_c, B_c, C_c, D_c) and the instrument precision $(\gamma_{a_i}, \gamma_{s_i})$, or else the algorithm proves infeasibility.

Step 3.   If the Economic Design Problem is infeasible, replace the last instrument (deleted on the previous iteration) and STOP, or return to Step 1 and increase one of the allowable upper bounds (on cost, control, or output). If Step 2 is feasible, continue to Step 4.

Step 4.   Delete the actuator or sensor that requires the least precision (largest among γ_{a_i} and γ_{s_i}). Return to Step 2.

10 Examples

The proposed sensor/actuator selection scheme is now applied to the control of a flexible structure. For flexible structures the question of actuator and sensor selection is of particular importance since such problems often require the implementation of a large number of sensors and actuators. This can make the sensor/actuator selection complex and expensive.

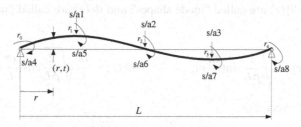

Fig. 1 Simply supported beam

Consider a simply supported beam [16] as in Figure 1. The beam has deflection $\mu(r,t)$ only in the plane of the paper, where r is the position from the left end of the beam. Consider five nodes $r_0, r_1, \ldots r_4$ evenly distributed on the beam. At nodes r_1, r_2, r_3, there are both force and torque actuators, collocated with displacement and velocity sensors. At the end nodes r_0, r_4, there are torque actuator collocated with velocity sensors. In model (1), such actuators are collected in the vector of inputs u and sensors are collected in the vector of outputs z. The outputs of interest, the vector y in the model (1), are the velocities at r_0, r_2, r_4. We consider an undamped flexible beam with N degrees of freedoms in the modal coordinates, q, modeled by the vector-second order differential equation

$$\ddot{q} + \Omega^2 q = H(u + w) \tag{18}$$

where

$$\Omega^2 = \mathrm{diag}(\omega_1^2, \ldots, \omega_N^2), \qquad H = \begin{bmatrix} H_1 & \ldots & H_N \end{bmatrix}^T,$$

where ω_i are the "modal frequencies", calculated by

$$\omega_i = \sqrt{\frac{EI}{\rho}} \left(\frac{i\pi}{L} \right)^2, \qquad i = 1, 2, \ldots, \infty,$$

EI is the modulus of elasticity, and ρ is mass density, both assumed constant. H_i are defined as follows:

For force inputs at r, $H_i = [\Psi_i(r_1), \Psi_i(r_2), \ldots]$;
For torque inputs at r, $H_i = [\Phi_i(r_1), \Phi_i(r_2), \ldots]$;

Table 1 Distribution of precisions.

	r_0	r_1	r_2	r_3	r_4
F		11.729	13.428	11.729	
T	21.370	12.836	13.434	12.836	21.370
q		2.195	13.356	2.195	
$\dot{q}$	1.544	0.475	0.802	0.475	1.544

If there is a force input at r_1 and a torque input at r_2, then $H_i = [\Psi_i(r_1), \Phi_i(r_2), \ldots]$. The functions $\Psi_i(r)$ are called "mode shapes" and $\Phi_i(r)$ are called "mode slopes". For $i = 1, 2, \ldots$,

$$\Psi_i(r) = \sqrt{\frac{2}{\rho L}} \sin(\frac{i\pi}{L} r), \qquad \Phi_i(r) = \frac{i\pi}{L} \sqrt{\frac{2}{\rho L}} \cos(\frac{i\pi}{L} r).$$

Define the state vector

$$x = \left[q_1, \frac{\dot{q}_1}{\omega_1}, \ldots, q_N, \frac{\dot{q}_N}{\omega_N} \right]^T$$

There are eight inputs $u = [u_1, \ldots, u_8]$, where u_1, u_2, u_3 are force inputs; $u_4, \ldots, u_8$ are torques. With eight sensors placed as in Figure 1, the coefficient matrix M can be determined accordingly. Then the state space model of the simply supported beam can be constructed in the form (1). For simplicity, let $N = 10$, $L = \pi$, $\rho = 2/L$, $EI = \rho$.

Next we will use the simply supported beam as an example to examine the relationships between the three competing factors; the control performance, control effort and instrument cost. First, let's fix the output covariance upper bound and the control covariance upper-bound, and then find the economic distribution of sensor/actuator precisions (minimize cost). Alternatively, we will fix the output requirements $\bar{Y}$ only, and find the relationship between the control effort and the instrument cost. After that we show how the instrument cost changes with the control performance requirements with fixed input effort upper bound. In the end, we show the instrument cost changes with both control performance and control effort. This provides some insight about information architecture in control design.

Given $\bar{Y}$ and $\bar{U}$, find $\mathbf{W}_{as}$

Consider the convex optimization problem (14). It can be converted to LMIs using Theorem 1. Hence it can be solved by any available LMI solvers. Let $\bar{Y} = \alpha I$, $\bar{U} = \beta I$, $\bar{\gamma}_a = \bar{\gamma}_s = \gamma$ for simplicity. When $\alpha = 1, \beta = 0.1, \gamma = 30$, the distribution of sensor/actuator precisions are shown in Table 1. It can be seen that the precision distribution is symmetric with respect to the center of the beam (and of course this is due to the symmetry of the required control performance). Among the 8 actuators, the torque actuators T at the end of the beam require the best precision, and hence the algorithm is suggesting that this actuator has great influence in achieving our objectives, and should not be deleted. Among the sensors, the displacement sensor q at the center of the beam requires the best precision. The corresponding output

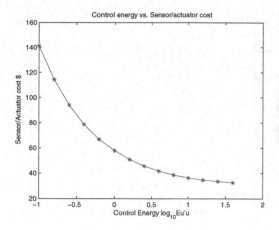

Fig. 2 $\alpha = 1, \beta \in [0.1, 40]$

feedback controller should have full order 20. Numerical calculations yield an almost singular matrix $Y^{-1} - X$, which indicates that the controller order require is less than full order. In this desirable situation, a reduced controller with order 15 is derived using the procedure described in [6].

Given $\bar{Y}$, find $\bar{U}$ vs. $

Fix $\alpha = 1$. For $\bar{U} = \beta I$, let β change from 0.1 to 40, while solving the optimization problem (14) for each $\bar{U}$. Figure 2 shows that when the control effort is big, we can preserve the same performance with less precise sensor/actuator. In other words, the control input and instrument can be seen as two types of costs. They are complementary to each other to maintain a certain performance.

Fix $\bar{U}$, find $\bar{Y}$ vs. $

Fix $\beta = 0.1$. Let α changes from 0.1 to 46 while solving the optimization problem (14) for each $\bar{Y}$. Figure 3 shows that the sensor/actuator cost and the performance are proportional to each other. It is reasonable that using less control energy will produce less output performance and less cost.

By plotting $\bar{U}$ vs. $\bar{Y}$ for a fixed γ, one will see the curve of maximal accuracy. Now we are ready to present the sensor/actuator cost vs a range of control inputs and output performance. Figure 4 illustrates the relationship of the three competing factors. This allows one to determine where to allocate resources in the early stage of control system design.

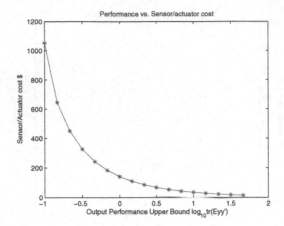

Fig. 3 $\alpha \in [0.1, 46], \beta = 0.1$

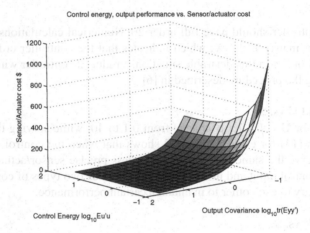

Fig. 4 $\alpha \in [0.1, 46], \beta \in [0.1, 40]$

11 Economic sensor/actuator selection

In the previous section, we have found the instrument precision requirements to achieve the input/output performance. Table 1 listed the necessary precision for each sensor/actuator. This is based on the assumption that sensor/actuator locations are known in a prior. Algorithm 1 proposes a systematic scheme to determine the desired location of sensors/actuators, the number and the precision of instruments required. This section is an application of Algorithm 1. In the first run, given in Table 1, we can see the velocity sensors at r_1 or r_3 require the minimum precision, so we delete one of them then solve the optimization problem (14) again.

Choose $\alpha = 0.001$, that is, specify the upper bound of the maximal accuracy $Y = 0.001I$; Let $\mu = 1$, namely, the price ratio between actuators and sensors is 1.

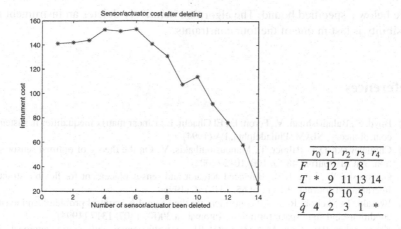

	r_0	r_1	r_2	r_3	r_4
F		12	7	8	
T	*	9	11	13	14
q		6	10	5	
$\dot{q}$	4	2	3	1	*

Fig. 5 Economic sensor/actuator selection

Apply Algorithm 1, recursively deleting the least important sensor/actuator until 4 sensors/actuators remain. The total sensor/actuator cost vs. the number of sensor/actuator is shown as in Figure 5, along with a table showing the deleting sequence. It can be seen that when $\mu = 1$, the sensors tend to be deleted more often, and among the 8 sensors, the rate sensors require less precision. Among the 8 actuators, the algorithm deleted force actuators. The torque actuator and the rate sensor at the beam require significantly greater precision than the other sensors/actuators, thus are most sensitive to the output performance.

12 Conclusion

The advances in this paper allow control theory to also decide the best information architecture. The selection of instrument precision (signal-to-noise ratio) is jointly determined with the control law, and this joint problem is shown to be convex. The problem solved is a feasibility problem with four constraints: max allowable instrument cost; max available instrument precision; max allowable output covariance; max allowable control covariance. Of course one can iteratively reduce a selected level set until feasibility is lost in any one of the four constraints, providing a minimal amount of resource of that variable (cost, output performance, or controller effort), in order to satisfy the others. So the same algorithm has four varieties for application. The optimal allocation of resources for state estimation problems also follows as a corollary of the main result.

The second contribution of the paper is to apply the above result in an *ad hoc* algorithm to reduce the number of sensors and actuators, from a large admissible set. The algorithm minimizes the weighted sum of the signal-to-noise ratios of all instruments, while choosing a controller that keeps the closed loop output covariance below a specified bound. The algorithm iteratively deletes an instrument until feasibility is lost in one of the four constraints.

References

[1] Boyd, S., Baladrishnan, V., Feron, E., El Ghaoui, L.: Linear matrix inequalities in system and control theory. SIAM, Philadelphia, PA (1994)

[2] Chmielewski, D.J., Palmer, T., Manousiouthakis, V.: On the theory of optimal sensor selection. AIChE Journal **48**(5), 100–1012 (2002)

[3] Gawronski, W., Lim, K.B.: Balanced actuator and sensor placement for flexible structures. International Journal of Control **65**, 131–145 (1996)

[4] Iwasaki, T., Skelton, R.E.: All controllers for the general h_∞ control problem: Lmi existence conditions and state space formulas. Automatica **30**(8), 1307–1317 (1994)

[5] Khargonekar, P.P., Rotea, M.A.: Mixed H_2/H_∞ control:a convex optimization approach. IEEE Transactions on Automatic Cotnrol **39**, 824C837 (1991)

[6] Li, F., de Oliveira, M. C., Skelton, R. E.: Integrating Information Architecture and Control or Estimation Design. SICE Journal of Control, Measurement, and System Integration **1**, 120–128 (2008)

[7] Lim, K.B.: Method for optimal actuator and sensor placement for large flexible structures. Journal of Guidance, Control and Dynamics **15**, 49–57 (1992)

[8] Liu, K., Skelton, R., Grigoriadis, K.: Optimal controllers for finite wordlength implementation. IEEE Transactions on Automatic Control **37**, 1294–1304 (1992)

[9] Lu, J.., Skelton, R.E.: Integrating instrumentation and control design. International journal of control **72**(9), 799–814 (1999)

[10] Lu, J., Skelton, R.: Economic design problem: Integrating instrumentation and control. In: Proc. American Control Conference, vol. 6, pp. 4321–4325. San Diego (1999)

[11] Mullis, C.T., Roberts, R.A.: Synthesis of minimum roundoff noise fixed point digital filters. IEEE Transactions on Circuit and System **23**(5), 256–262 (1976)

[12] Naeije, W.J., Bosgra, O.H.: The design of dynamic compensators for linear multivariable systems. In: Proc. 4th IFAC Symposium on Multivariable Technological Systems, pp. 205–213. Fredericton, Canada (1977)

[13] Oliveira, M., J.C.Geromel: Linear output feedback controller design with joint selection of sensors and actuators. IEEE Transaction on Automatic Control **45**(12), 2412–2419 (2000)

[14] de Oliveira, M.C., Geromel, J.C., Bernussou, J.: Extended $\mathbf{H}_2$ and $\mathbf{H}_\infty$ norm characterizations and controller parametrizations for discrete-time systems. International Journal of Control **75**(9), 666C679 (2002)

[15] Scherer, C., Gahinet, P., Chilali, M.: Mulitobjective output-feedback cotnrol via lmi optimization. IEEE Transactions on Automatic Control **42**(7), 896–911 (1997)

[16] Skelton, R.E.: Dynamic systems control: linear systems analysis and synthesis. John Wiley & Sons, New York (1988)

[17] Skelton, R.E., Delorenzo, M.L.: Space structure control design by variance assignment. Journal of guidance and control **8**, 454–462 (1985)

[18] Skelton, R.E., Iwasaki, T., Grigoriadis, K.: A unified algebraic approach to linear control design. Taylor & Francis, London (1998)

[19] Skelton, R.E., Li, F.: Economic sensor actuator selection and its application in structure control. In: SPIE International Symposium on Smart Structures. San Diego,USA (2004)

[20] Sripad, A., Snyder, D.: A necessary and sufficient condition for quantization error to be uniform and white. IEEE Transactions on Acoustics, Speech and Signal Processing **5**(10), 442C448 (1977)

[21] Williamson, D., Kadiman, K.: Optimal finite wordlength linear quadratic regulation. IEEE transactions on automatic control **34**(12), 1218–1228 (1989)

Uncertain Model Set Calculation from Frequency Domain Data

Gary J. Balas, Andrew K. Packard and Peter J. Seiler

Abstract Uncertainty models play a central role in the robust control framework. The uncertainty models and their structure determine the design trade off between performance and robustness of the closed-loop system. Therefore, given a nominal multivariable model of the system, a set of multivariable frequency response measurements, and model structure, it is desirable to generate a corresponding model set which tightly over bounds the given data. We show that computation of the model set for a given structure, which is consistent with the data, can be formulated as a linear matrix inequality feasibility problem. Formulas are derived which allow comparison between model structures to assess the relative size of the each model set. The proposed algorithms are applied to the lateral-directional axis of a radio-controlled aircraft developed by NASA Langley researchers.

1 Introduction

This paper considers algorithms to optimally cover finite sets of matrices with diagonally-weighted unit balls (with Euclidean operator norm) of matrices. The problem is reformulated as an semidefinite program. The resultant ball is always a subset of the smallest scalar-weighted ball which covers the finite set, and an easily computed measure which quantifies the degree of containment is developed.

Gary J. Balas
Aerospace Engineering and Mechanics, University of Minnesota, Minneapolis, MN U.S.A 55455, e-mail: balas@umn.edu

Andrew K. Packard
Mechanical Engineering, University of California, Berkeley, CA U.S.A 94720, e-mail: apackard@berkeley.edu

Peter J. Seiler
Aerospace Engineering and Mechanics, University of Minnesota, Minneapolis, MN U.S.A 55455, e-mail: seiler@aem.umn.edu

P.M.J. Van den Hof et al. (eds.), *Model-Based Control: Bridging Rigorous Theory and Advanced Technology*, DOI: 10.1007/978-1-4419-0895-7_6,
© Springer Science + Business Media, LLC 2009

These results are applied frequency-by-frequency to obtain uncertainty models from frequency response data. The approach is similar to past work by Hindi, et. al [1]. The recent work by Volker and Engell [4] and Haggblom [3] are also relevant. The motivation for this problem is multi-model robust control. Often, one has a finite number of plants for which a robust controller must be designed. These plants may represent a single physical system at different operating points, e.g. an aircraft at different trim conditions. Alternatively they may represent the variations in a set of similar physical systems, such as a collection of hard disk drives. In either case, it is known that the simultaneous stabilization problem of two plants by a single controller can be reformulated into a strong stability problem for which a pole-zero interlacing condition is necessary and sufficient [2]. The existence of a controller which simultaneously stabilizes 3 plants is an undecidable problem [5]. A relaxation of the multi-model design problem is to cover the finite set of plants with an additive or multiplicative uncertainty model and then design a controller which stabilizes all models in the uncertainty set. This relaxation can be turned into an H_∞ design for the case where the uncertainty model contains only a single unstructured uncertainty.

2 Uncertainty Models

$\mathscr{R}$ denotes proper, rational functions, and $\mathscr{R}^{n \times m}$ is the set of $n \times m$ matrices with elements in $\mathscr{R}$. Similarly $\mathscr{S}$ denotes stable (poles in open left-half plane), proper, rational functions. For any ring $\mathscr{K}$, $\mathscr{U}_{\mathscr{K}^{n \times n}}$ denotes the units in $\mathscr{K}^{n \times n}$. For $G \in \mathscr{R}^{n \times m}$, the notation $U_{\#}(G)$ is the number of unstable eigenvalues in a stabilizable and detectable realization of G.

Given $P \in \mathscr{R}^{n \times m}$, and $W_L \in \mathscr{U}_{\mathscr{S}^{n \times n}}, W_R \in \mathscr{U}_{\mathscr{S}^{m \times m}}$, define 3 different sets of systems, each representing perturbations of P

$$\mathscr{A}(P, W_L, W_R) := \left\{ \tilde{P} = P + W_L \Delta W_R : \Delta \in \mathscr{R}^{n \times m}, \|\Delta\|_{\mathscr{L}_\infty} \leq 1, U_{\#}(P) = U_{\#}(\tilde{P}) \right\}$$

$$\mathscr{M}_{\mathscr{I}}(P, W_L, W_R) := \left\{ \tilde{P} = P(I_m + W_L \Delta W_R) : \Delta \in \mathscr{R}^{n \times m}, \|\Delta\|_{\mathscr{L}_\infty} \leq 1, U_{\#}(P) = U_{\#}(\tilde{P}) \right\}$$

$$\mathscr{M}_{\mathscr{O}}(P, W_L, W_R) := \left\{ \tilde{P} = (I_n + W_L \Delta W_R)P : \Delta \in \mathscr{R}^{n \times m}, \|\Delta\|_{\mathscr{L}_\infty} \leq 1, U_{\#}(P) = U_{\#}(\tilde{P}) \right\}$$

Theorem 1 *Given* $P \in \mathscr{R}^{n \times m}$, *and* $W_L \in \mathscr{U}_{\mathscr{S}^{n \times n}}, W_R \in \mathscr{U}_{\mathscr{S}^{m \times m}}$. *Assume P has no poles on the imaginary axis. Then* $\tilde{P} \in \mathscr{R}^{n \times m}$ *satisfies* $\tilde{P} \in \mathscr{A}(P, W_L, W_R)$ *if and only if* $U_{\#}(P) = U_{\#}(\tilde{P})$ *and*

$$\begin{bmatrix} W_L W_L^* & \tilde{P} - P \\ (\tilde{P} - P)^* & W_R^* W_R \end{bmatrix} \succeq 0 \tag{1}$$

for all $\omega \in \mathbf{R}$.

Proof:

$\rightarrow$ By definition, there exists a $\Delta \in \mathscr{R}\mathscr{L}_\infty$ such that $\|\Delta\|_\infty \leq 1$ and $\tilde{P} = P + W_L \Delta W_R$. Consequently $\tilde{P}$ has no poles on imaginary axis. The weights are invertible everywhere in the right-half plane, so for each ω,

$$\Delta(j\omega) = W_L^{-1}(j\omega)\left[\tilde{P}(j\omega) - P(j\omega)\right]W_R^{-1}(j\omega) \quad \text{and} \quad \bar{\sigma}[\Delta(j\omega)] \leq 1.$$

This implies that for all ω, $\bar{\sigma}\left[W_L^{-1}(j\omega)\left[\tilde{P}(j\omega) - P(j\omega)\right]W_R^{-1}(j\omega)\right] \leq 1$ which is equivalent to equation (1).

$\leftarrow$ If $\tilde{P}$ has any poles on imaginary axis, then at frequencies near these poles, $\tilde{P}$ has elements with arbitrarily large magnitude. Since P, W_L and W_R are bounded on imaginary axis, the LMI in 1 cannot be feasible for all ω. Consequently $\tilde{P}$ has no poles on the imaginary axis. Define $\Delta := W_L^{-1}\left(\tilde{P} - P\right)W_R^{-1}$. Clearly $\Delta \in \mathcal{RL}_\infty$ and $\tilde{P} = P + W_L\Delta W_R$. Constraint (1) implies that for all ω, $\bar{\sigma}[\Delta(j\omega)] \leq 1$. Finally, with the additional assumption that $\tilde{P}$ has the same number of RHP poles as P, $\tilde{P} \in \mathscr{A}(P, W_L, W_R)$ follows.

In order to address multiplicative uncertainty models, we recall a simple theorem about minimum-norm solutions to linear matrix equations.

Theorem 2 *Suppose* $C, D \in \mathbf{C}^{n \times m}$ *with* $n \leq m$ *and* $\mathrm{rank}(C) = n$. *Then*

$$\min_{X \in \mathbf{C}^{m \times m}, CX = D} \bar{\sigma}(X) = \bar{\sigma}\left[(CC^*)^{-\frac{1}{2}}D\right]$$

Without loss in generality, we only consider the input-multiplicative case. Take $P \in \mathscr{R}^{n \times m}$, with no poles on imaginary axis, $n \leq m$. Assume that $P(j\omega)$ full row rank for all ω, so that any matrix $G \in \mathbf{C}^{n \times m}$ is in the range space of $P(j\omega)$, and hence multiplicative input uncertainty can transform $P(j\omega)$ into G.

Theorem 3 *Given* $P \in \mathscr{R}^{n \times m}$, *and* $W_L \in \mathscr{U}_{\mathscr{S}^{n \times n}}, W_R \in \mathscr{U}_{\mathscr{S}^{m \times m}}$. *Assume* P *has no poles on the imaginary axis. Then* $\tilde{P} \in \mathscr{R}^{n \times m}$ *satisfies* $\tilde{P} \in \mathscr{M}_{\mathscr{S}}(P, W_L, W_R)$ *if and only if* $U_\#(P) = U_\#(\tilde{P})$ *and*

$$\begin{bmatrix} PW_LW_L^*P^* & \tilde{P} - P \\ (\tilde{P} - P)^* & W_R^*W_R \end{bmatrix} \succeq 0 \tag{2}$$

for all $\omega \in \mathbf{R}$.

Proof:

$\rightarrow$ By definition, $\exists \Delta \in \mathscr{RL}_\infty$ such that $\|\Delta\|_\infty \leq 1$ and $\tilde{P} = P(I_m + W_L\Delta W_R)$. Consequently $\tilde{P}$ has no poles on imaginary axis,, and for each ω,

$$P(j\omega)W_L(j\omega)\Delta(j\omega) = \left[\tilde{P}(j\omega) - P(j\omega)\right]W_R^{-1}(j\omega), \quad \text{and} \quad \bar{\sigma}[\Delta(j\omega)] \leq 1.$$

This implies that the minimum-norm solution (of X) to

$$P(j\omega)W_L(j\omega)X = \left[\tilde{P}(j\omega) - P(j\omega)\right]W_R^{-1}(j\omega)$$

satisfies $\bar{\sigma}(X) \leq 1$, which is equivalent to equation (2).

←   Since (2) holds, $\tilde{P}$ cannot have any imaginary axis poles. Define

$$\Delta := (PW_L)^* \, (PW_L W_L^* P^*)^{-1} \, (\tilde{P} - P) \, W_R^{-1}.$$

The full row-rank assumption on P implies that $\Delta \in \mathscr{RL}_\infty$, and substitution shows $\tilde{P} = P(I_m + W_L \Delta W_R)$. Constraint (2) implies that for all ω, $\bar{\sigma}\,[\Delta(j\omega)] \leq 1$. Consequently, Δ is indeed proper, and $\|\Delta\|_\infty \leq 1$. Finally, since $U_\#(P) = U_\#(\tilde{P})$, $\tilde{P} \in \mathscr{M}_\mathscr{I}(P, W_L, W_R)$ as desired.

Other uncertain models can also be represented as solutions to LMI conditions. For example, a single-block uncertainty model incorporating both additive and multiplicative can be expressed in a similar manner. Since the representation includes an additive term, there are no dimension and/or rank constraints on P. This will be presented in a later paper.

2.1 Application to covering a family of models

One motivation for this paper is multi-model robust control. Through modeling, or sampling, or experimentation, one has a finite number of plants for which a robust controller must be designed. A relaxation of the multi-model design problem simply covers the finite set of plants with an additive or multiplicative uncertainty model, and bases the design on the cover.

Given a nominal model $P \in \mathscr{R}^{n \times m}$, and a finite collection of models $\{P_i\}_{i=1}^N$, how can weights W_L and W_R be chosen so that for all i

$$P_i \in \mathscr{A}\,(P, W_L, W_R) \quad \text{or} \quad P_i \in \mathscr{M}_I(P, W_L, W_R) \quad \text{or} \quad P_i \in \mathscr{M}_O(P, W_L, W_R)$$

By assumption, one must first assume, or verify, that all models have the same number of right-half plane poles, since this is part of the uncertain sets' definitions. Then, the weights W are chosen so that the LMI's hold at all frequencies.

The goal is to pick W_L and W_R as "small" as possible, since these roughly represent a "radius" of the uncertain set. Proceeding informally, define new optimization variables L and R for $W_L W_L^*$ and $W_R^* W_R$ respectively. A plausible optimization to determine small, feasible weights is:

1) Chose positive definite matrices Γ_1 and Γ_2
2) Pick a nominal plant model, P
3) Grid the frequency axis using $\{\omega_k\}_{k=1}^M$.
4) At each frequency, solve the SDP

$$\min_{L \succ 0, R \succ 0} \operatorname{Tr} \Gamma_1 L + \operatorname{Tr} \Gamma_2 R$$

subject to

$$\begin{bmatrix} L & P_i - P \\ (P_i - P)^* & R \end{bmatrix} \succeq 0 \quad \text{or} \quad \begin{bmatrix} PLP^* & P_i - P \\ (P_i - P)^* & R \end{bmatrix} \succeq 0$$

for all $\{i\}_{i=1}^N$, and at all frequencies $\{\omega_k\}_{k=1}^M$.

The positive-definite matrices Γ_i, which are used to weight the objective function c can also be functions of frequency.

If the uncertain models are going to be used in a robust synthesis procedure, it may be necessary to obtain rational functions instead of a set of values at a finite set of frequencies. In this situation, it is practical to consider diagonal W_L and W_R (resulting in real, diagonal L and R). Then (element-wise) we can use log-Chebychev fitting techniques, yielding a stable, minimum-phase transfer function whose magnitude over-bounds the optimal frequency-by-frequency values.

In either case, we have no particular rules by which to choose the Γ weightings in the cost function. Different choices will lead to different optimal W, and hence different coverings of the data, $\{P_i\}_{i=1}^N$. In the next section, we introduce a measure to compare to coverings of the same data.

2.2 Containment Metrics

We focus attention to a constant matrix problem. Given complex matrices $C \in \mathbf{C}^{n \times m}, D \in \mathbf{C}^{n \times n}, E \in \mathbf{C}^{n \times m}$, define

$$\mathscr{B}_\alpha(C,D,E) := \{C + D\Delta E : \Delta \in \mathbf{C}^{n \times n}, \bar{\sigma}(\Delta) \le \alpha\}$$

When $\alpha = 1$, simply write $\mathscr{B}(C,D,E)$.

Given two such triples (C_1,D_1,E_1) and (C_2,D_2,E_2), what is the smallest value of α such $\mathscr{B}_\alpha(C_1,D_1,E_1)$ contains $\mathscr{B}(C_2,D_2,E_2)$? How is this motivated? Suppose $\mathscr{B}(C_1,D_1,E_1)$ and $\mathscr{B}(C_2,D_2,E_2)$ are two different "covers" of a finite collection of matrices. In order to gain insight over which might be a less conservative cover, we ask 2 questions: How much does the set described by (C_1,D_1,E_1) have to be expanded to include all of $\mathscr{B}(C_2,D_2,E_2)$, and conversely, how much does the set described by (C_2,D_2,E_2) have to be expanded to include all of $\mathscr{B}(C_1,D_1,E_1)$? Generally, both questions will give different answers, and it is intuitive that the description which has to be expanded more is a less conservative cover. These are considered containment metrics. Unfortunately, this notion is not transitive, so the results of such an analysis are meant to be insightful, not infallible.

Theorem 4 *Theorem: Suppose $n \ge m$, $C_i \in \mathbf{C}^{n \times m}$, $D_i \in \mathbf{C}^{n \times n}$, $E_i \in \mathbf{C}^{n \times m}$, with each D_i invertible, and each E_i full column rank. Then*

$$\min\{\alpha : \mathscr{B}(C_2,D_2,E_2) \subset \mathscr{B}_\alpha(C_1,D_1,E_1)\} = \max_{\Delta_2 \in \mathbf{C}^{n \times n}, \bar{\sigma}(\Delta_2) \le 1} \bar{\sigma}[M(\Delta_2)]$$

where $M(\Delta) = D_1^{-1}(C_2 - C_1 + D_2 \Delta E_2)(E_1^* E_1)^{-\frac{1}{2}}$ Moreover, the maximization can be computed reliably and accurately with structured singular value/SDP methods.

Remark: An analogous result holds for E_i invertible, and D_i full row rank.

Proof: For any $\Delta_2 \in \mathbf{C}^{n \times n}$, a minimum norm Δ_1 solving

$$C_1 + D_1 \Delta_1 E_1 = C_2 + D_2 \Delta_2 E_2$$

is given by

$$D_1^{-1}(C_2 - C_1 + D_2 \Delta_2 E_2)(E_1^* E_1)^{-\frac{1}{2}}$$

Therefore, for any $\alpha > 0$, $\mathscr{B}(C_2, D_2, E_2) \subset \mathscr{B}_\alpha(C_1, D_1, E_1)$ if and only if

$$\max_{\Delta_2 \in \mathbf{C}^{n \times n}, \bar{\sigma}(\Delta_2) \leq 1} \bar{\sigma}\left[D_1^{-1}(C_2 - C_1 + D_2 \Delta_2 E_2)(E_1^* E_1)^{-\frac{1}{2}}\right] \leq \alpha$$

A few special cases can be solved analytically. For example, consider an input-multiplicative model $P(I + W_1 \Delta_1 W_2)$, and an additive model $P + W_3 \Delta_2 W_4$, with P square and invertible. Using the terminology of section 2, the nominal plant model is the same for both uncertain representations, and is square, and invertible. Then

- the input-multiplicative model must be increased by a factor $\bar{\sigma}\left[(PW_1)^{-1} W_3\right] \bar{\sigma}\left[W_4 W_2^{-1}\right]$ in order to contain the additive model; and
- the additive model must be increased by a factor $\bar{\sigma}\left[W_3^{-1}(PW_1)\right] \bar{\sigma}\left[W_2 W_4^{-1}\right]$ in order to contain the input-multiplicative model.

These quantities will be used in section 3.5 to compare different uncertain representations.

3 Application of Over-Bound Uncertainty Modeling to NASA GTM Aircraft

Over-bounding techniques are used to generate input multiplicative and additive uncertainty models for the NASA Generic Transport Model (GTM) aircraft, see Figure 1 [6, 7]. The GTM is a 5.5% dynamically-scaled, remotely piloted, twin-turbine swept wing aircraft developed as part of the Airborne Sub-scale Transport Aircraft Research (AirSTAR) test-bed at NASA Langley Research Center. The GTM aircraft will be used for research experiments pertaining to dynamics modeling and control beyond the normal flight envelope.

3.1 Lateral-Directional GTM Aircraft Linear Model

A lateral-directional linear model of the GTM aircraft, **LAT**, is derived for straight and level flight at an airspeed of 40 m/s and an altitude of 200 m. The states are side slip (β, rad), roll rate (p, rad/s), yaw rate (r, rad/s) and bank angle (ϕ, rad), inputs to the model are angular deflections of the ailerons (δ_{ail}, rad) and rudder (δ_{rud}, rad) and the outputs are side slip, roll rate, and yaw rate. The state-space linearized model is shown in Table 1. Bode plots of these transfer functions are shown in Figure 2.

$$
\mathbf{LAT} = \left[
\begin{array}{cccc|cc}
-0.593 & 9.997 & -129.151 & 32.082 & -1.394 & 26.015 \\
-0.644 & -5.731 & 1.938 & 0 & -47.045 & 15.599 \\
0.232 & -0.276 & -1.446 & 0 & -1.372 & -24.846 \\
0 & 1 & 0.076 & 0 & 0 & 0 \\
\hline
0.008 & 0 & 0 & 0 & 0 & 0 \\
0 & 1 & 0 & 0 & 0 & 0 \\
0 & 0 & 1 & 0 & 0 & 0
\end{array}
\right]
$$

Table 1 Linearized GTM model at 40 m/s, 200 m altitude

3.2 Generation of Frequency Response Data Sets

The uncertain, lateral-directional GTM model shown in Figure 3 is used to generate several sets of input/output frequency response data. This uncertain model includes parametric uncertainty of 4% in the GTM stability derivatives C_{L_β}, C_{Y_β}, C_{L_p}, C_{L_r}, and C_{N_p} and 8% in the stability derivatives C_{Y_β} and C_{N_r}. Δ_{stab} denotes the normalized block diagonal uncertainty corresponding to these parametric uncertainty. In addition to uncertainty in the stability derivatives, unmodeled dynamic is included as input multiplicative uncertainty and sensor noise is included with the weight W_{sn}. This is represented in Figure 3.

The input multiplicative uncertainty weight in the aileron channel, $W_{ail} = \frac{4s+4.85}{s+97}$. It has a low frequency gain of 0.05, crosses 1 at 25 rad/s, and a high frequency

Fig. 1 Generic Transport Model Aircraft

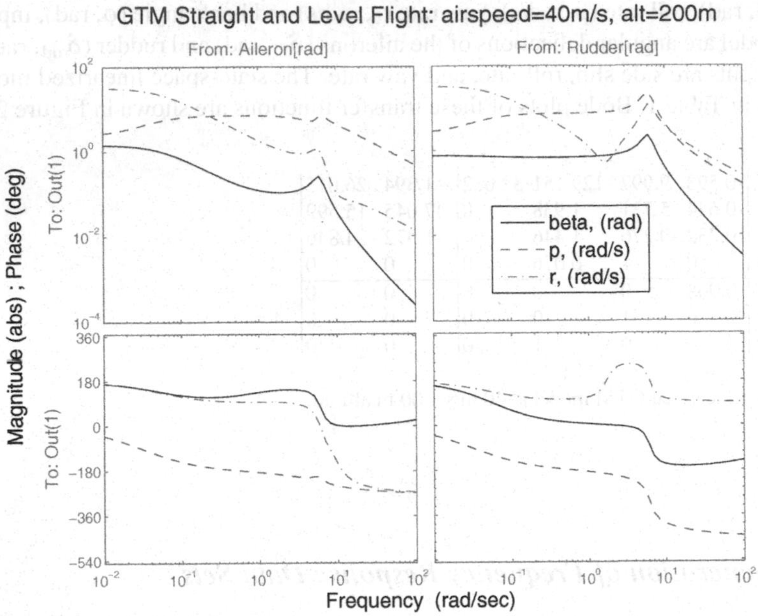

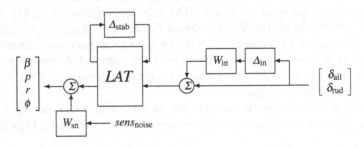

GTM Straight and Level Flight: airspeed=40m/s, alt=200m

Fig. 2 Bode Plot from Inputs $[\delta_{\text{ail}}, \delta_{\text{rud}}]$ to Outputs $[\beta, p, r]^T$

Fig. 3 Uncertain Model GTM Model

gain of 4. The input multiplicative uncertainty weight in the rudder channel is $W_{\text{rud}} = \frac{4s+9.73}{s+97.3}$. It has a low frequency gain of 0.10, crosses 1 at 25 rad/s, and a high frequency gain of 4. The weight W_{in} in Figure 3 denotes 2×2 system with W_{ail} and W_{rud} on the diagonals, $W_{\text{in}} = diag(W_{\text{ail}}, W_{\text{rud}})$. The sensor noise is modeled with a scalar gain in each channel of 0.05, $W_{\text{sn}} = 0.05 \cdot I_{3 \times 3}$.

Three sets of input/output frequency response data are generated to examine the properties of the input multiplicative and additive uncertainty over-bound descriptions. The data sets are denoted I for input, IP for input and parametric, and IPN for input, parametric, and sensor noise. In generating data set I, the sensor noise and Δ_{stab} are set to zero. Input-output frequency response data is then generated by randomly sampling Δ_{in}. Similarly, data set IPN is generated using both uncertainty descriptions and the sensor noise shown in Figure 3. Twenty five input/output frequency responses are generated for each data set.

One point should be clarified before proceeding. The uncertainty model used to generate the I, IP, and IPN data sets should be viewed as the "true" uncertainty model. In many real applications, input/output frequency response data is obtained from experimental measurements and the form of the "true" uncertainty model generating this data is not available. In such cases, simple uncertainty models such as input multiplicative and additive uncertainty are commonly used to over-bound the frequency response data sets. Even if the form of the "true" uncertainty model is known, a simple uncertainty over-bound might be used since it can simplify the control design. In the following sections, the frequency responses in data sets I, IP, and IPN will be covered with various combinations of input multiplicative and additive uncertainty. The examples will demonstrate the issues which can arise when these simple uncertainty models are used to cover frequency data generated from more complicated "true" uncertainty models.

3.3 Over-Bounding as a LMI Feasibility Problem

Given a nominal model and a collection of frequency domain data, an input multiplicative or additive uncertainty over-bound weight at each frequency point can be developed by solving a linear, matrix inequality. The nominal linear model must be full rank at each frequency point for there to exist a solution to the linear, matrix inequality.

Initially a two-input, two-output model whose outputs are side slip, β, and roll rate, p is used for analysis. The singular values of the two-input, two-output lateral-directional GTM model, inputs δ_{ail} and δ_{rud} and outputs β and p, are plotted in the frequency range of interest, 0.01 to 100 rad/s in Figure 4. The nominal model is full rank across the frequency range of interest. There is approximately a factor of 20 difference between the two singular values which indicates that it is easier to control the aircraft in one direction than the other. This will have implications on the magnitude of the over-bound weights required to describe the difference between the nominal and other 25 models in the set.

3.3.1 Data Set I

Consider generation of over-bound weights for data set I. Recall that data set I was generated from a "true" uncertainty model consisting of the GTM linear model with only input multiplicative uncertainty. Thus one would expect that using a input multiplicative uncertainty model to over bound the error should exactly reproduce the true uncertainty model. For comparison, an additive uncertainty over bound is generated for the same data set. A Bode plot of the nominal GTM aircraft and model

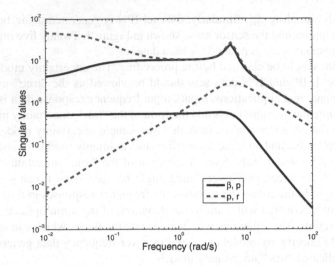

Fig. 4 Singular Values

data set are shown in Figure 5. The nominal model transfer functions are marked
with +.

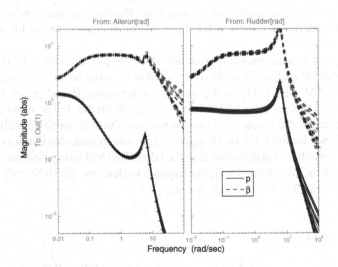

Fig. 5 Bode Plot of Data Set I

Input multiplicative, $P_{\text{in}} = P(I + W_1\Delta_{\text{in}}W_2)$, and additive, $P_{\text{add}} = P + W_3\Delta_{\text{add}}W_4$,
uncertainty over-bound models of the error between the nominal lateral-directional
GTM model and the frequency response data are generated by solving the linear ma-

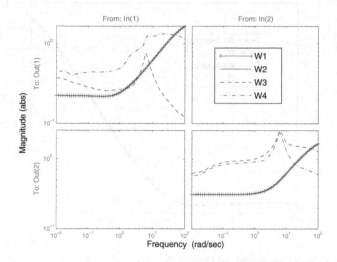

Fig. 6 Input Multiplicative and Additive Over-Bound Weights

trix inequalities (LMIs) in equations 1 and 2 at each frequency point. The LMI optimization searches for the smallest magnitude frequency varying weights W_1 and W_2 which account for all the model error. The input multiplicative over-bound weights W_1 and W_2 are identical and the product of their diagonal entries is almost exactly equal to the original weights W_{ail} and W_{rud}. The additive uncertainty weights, W_3 and W_4, vary significantly across frequency to capture to plant variation due to the input multiplicative uncertainty as seen in Figure 6.

An input multiplicative over-bound model of the error between the nominal lateral-directional GTM model and the model set can be described in the same format as the true model description, $P(I + W_1 \Delta_{in})$, by setting W_2 to be the identity matrix. This structure is used in the subsequent input multiplicative over-bound investigations for data sets IP and IPN. The input multiplicative over bound weight W_1 is nearly identical to the original weights W_{ail} and W_{rud} as seen Figure 7.

3.3.2 Data Set IP

Consider the case of generating over-bound uncertainty models for data set IP. The true uncertainty model used to generate data in the set IP includes GTM models whose stability derivatives are allowed to vary between 4% and 8%. Since the uncertainty in the stability derivatives is small, one would expect an uncertainty model using input multiplicative uncertainty to provide a tight over-bound on the frequency response data. An additive uncertainty over bound is derived for the same data set for comparison. A Bode plot of the nominal GTM aircraft and model set data is shown in Figure 8.

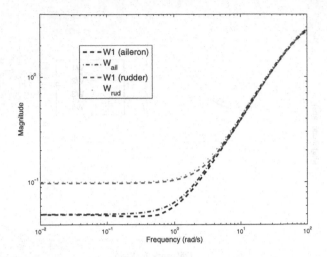

Fig. 7 Comparison of Over-Bound and Original Weights

Input multiplicative and additive uncertainty over-bound models of the error are derived and presented in Figure 9. They are plotted on the graphs to conserve space. The input multiplicative uncertainty model is able to capture the difference between the nominal model and frequency response data. The weight, W_1, matches the input multiplicative weight in the true uncertainty models above 1 rad/s as seen in Figure 9. Below 1 rad/s, the variations of the stability derivatives in the true uncertainty model result in the input multiplicative over-bound weight being larger at low frequencies. These same characteristics are evident from the additive uncertainty over bound results (Figure 9).

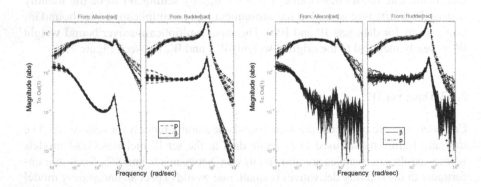

Fig. 8 Bode Plot of Data Set IP (left) and Data Set IPN (right)

3.3.3 Data Set IPN

Consider the case of over-bounding the data in data set IPN. A Bode plot of the nominal GTM aircraft and IPN model set data are shown in Figure 8. The sensor noise weight in the true uncertainty model is relatively small, 0.05, when compared with the magnitude of the nominal transfer functions above 1 and below 20 rad/s. The input multiplicative and additive over-bound weights are shown in Figure 10 for the IPN data set. As before they are plotted on the same graph to conserve space. Note that the input multiplicative over-bound weights are slightly larger in magnitude in the mid-frequency range, 0.1 to 5 rad/s, and above 10 rad/s relative to the previous two sets of frequency response data. In the mid-frequency range, where the ratio of the maximum and minimum singular values is about 20, one direction dominates in the nominal model. The sensor noise in the true uncertainty model is distributed equally between the channels, hence to over-bound the "round" sensor noise the input multiplicative uncertainty weights must be increased.

The sensor noise is agnostic in its favoring of one direction or another. Hence the additive uncertainty weights are minimally effected by the addition of sensor noise as is seen in Figure 10.

3.4 Effect of System Directionality

The results of the previous sections showed the important role directionality plays in development of uncertainty over-bounds. Consider the lateral-directional model of

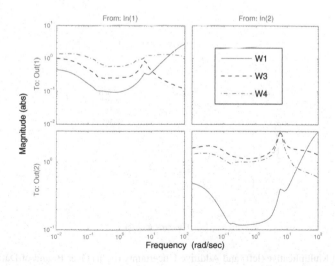

Fig. 9 Input Multiplicative and Additive Uncertainty Over-Bounds for Data Set IP

the GTM aircraft with the side slip (β, rad) output replaced by yaw rate (r, rad/s). A
Bode plot of the these transfer functions are shown in Figure 2. The singular values
of the new model are also plotted in the frequency range of interest (Figure 4). Note
that below 0.1 rad/s there is approximately a factor of 500 difference between the
two singular values. This indicates that at low frequency, the actuator inputs are able
to only control the aircraft in one direction. This has implications on the magnitude
of the input multiplicative over-bound weights required to describe the difference
between the nominal and other 25 responses in the set.

Consider data from set IPN. An input multiplicative over-bound weight is derived
for the GTM with roll rate and yaw rate outputs. The input multiplicative over-bound
weight W_1 for the roll rate and yaw rate output case are compared with the beta and
roll rate model. The input multiplicative over-bound weights for the roll rate and
yaw rate model are a factor of 50 larger than the same weights for beta and roll rate
model (Figure 11). This large difference is due to dominance of one direction in the
GTM with roll rate and yaw rate outputs. Hence it is important to understand the
limitations the nominal model dynamics may impose on the over-bound uncertainty
weights.

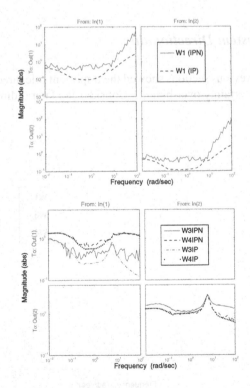

Fig. 10 Input Multiplicative (left) and Additive Uncertainty (right) Over-Bound of Data Set IPN

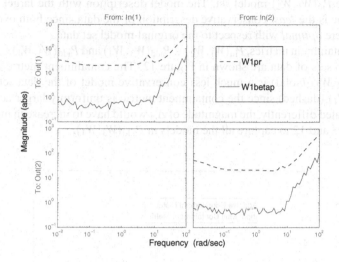

Fig. 11 Input Multiplicative Over-Bound Weight W_1

3.5 Containment Metric

Two uncertain model descriptions, $P_{in} = P_{nom}(I + W_1\Delta_{in}W_2)$ and $P_{add} = P_{nom} + W_3\Delta_{add}W_4$, were used to over-bound the same 25 frequency responses. It is of interest to compute the containment metric to compare the conservatism of these relative model sets. Note that for each model description, the over-bounding weights were optimal.

The input multiplicative and additive over-bound models will be compared for data case IP. Recall the containment metric determines how large, in magnitude, the model uncertainty bound (Δ_{in}) needs to grow to contain all the models described by the other (Δ_{add}) model description. The containment metric for the model defined by the additive uncertainty over-bound weights relative to the input multiplicative over-bound model, $P_{add}(W_3, W_4) \in P_{in}(W_1, W_2)$, is at each frequency ω.

$$|\Delta_{add}| = \bar{\sigma}(W_3^{-1}P_{nom}W_1)\bar{\sigma}(W_2W_4^{-1}), \qquad \text{at } \omega$$

The containment metric for the model defined by the input multiplicative uncertainty over-bound weights relative to the additive over-bound model, $P_{in}(W_1, W_2) \in P_{add}(W_3, W_4)$, is

$$|\Delta_{in}| = \bar{\sigma}((P_{nom}W_1)^{-1}W_3)\bar{\sigma}(W_4W_2^{-1}), \qquad \text{at } \omega$$

The containment metric would be 1 at a frequency if the model set, $P_{add}(W_3, W_4)$ for example, contains all the models in $P_{in}(W_1, W_2)$. The larger the containment metric, the larger the uncertainty magnitude in $P_{add}(W_3, W_4)$ would need to grow to contain the $P_{in}(W_1, W_2)$ model set. The model description with the larger containment metric is the *least* conservative description of the data since both over-bound weights were *optimal* with respect to the original model set data.

The containment metrics, $P_{in}(W_1, W_2) \in P_{add}(W_3, W_4)$ and $P_{add}(W_3, W_4) \in P_{in}(W_1, W_2)$, for the two sets of data are shown in Figure 12. The containment metric indicates that $P_{in}(W_1, W_2)$ (solid) is a much less conservative model of the data set IP than $P_{add}(W_3, W_4)$ (dashed) since the containment metric is uniformly larger across frequency. Stated differently, the magnitude of Δ_{in} would have to increase in magnitude between a 3 and 12 to capture all the models in $P_{add}(W_3, W_4)$.

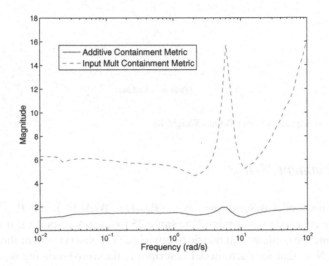

Fig. 12 Containment Metrics for Additive and Input Multiplicative Over-Bound Models

4 Conclusions

Algorithms were developed in this paper to optimally cover finite sets of matrices with diagonally-weighted unit balls (with Euclidean operator norm) of matrices. The problem was reformulated as a linear matrix inequality feasibility problem. Metrics were also derived to to assess the relative size of the each model set relative to a specific model structure. The proposed algorithms are applied to the lateral-directional axis of a radio-controlled aircraft and the results discussed.

Acknowledgements This research was partially supported by NASA Langley Research Center under SBIR NNL06AA03C entitled "Real Time Monitoring and Test-Vector Generation for Improved Flight Safety." Mr. Alec Bateman is the program manager at Barron Associates Inc. and Dr. Christine M. Belcastro is the technical monitor.

References

[1] Hindi, H., Seong, C-Y, and Boyd, S., Computing optimal uncertainty models from frequency domain data. Proceeding 41st IEEE Conf. Decision and Control, vol. 3, 2898–2905 (2002).

[2] Vidyasagar, M., Control System Synthesis: A Factorization Approach. MIT Press, 1985.

[3] Haggblom, M., Data-based modeling of block diagonal uncertainty by convex optimization. Proceedings of the American Control Conference, 4637-4642, 2007.

[4] Volker, M., and Engell, S., Computation of tight uncertainty bounds from time-domain data with application to a reactive distillation column. Proceedings of the IEEE Conference on Decision, 2939-2944, 2005.

[5] Blondel, V., and Gevers, M., Simultaneous stabilizability of three linear systems is rationally undecidable. Mathematics of Control, Signals, and Systems, vol. 6, no. 2, 135-145, 1993.

[6] Jordan, T., Langford, W, and Hill, J., Airborne subscale transport aircraft research testbed - Aircraft model development. AIAA Guidance, Navigation and Control Conf., AIAA-2005-6432 (2005).

[7] Jordan, T., Foster, J.V., Bailey, R.M, and Belcastro, C.M., AirSTAR: A UAV platform for flight dynamics and control system testing. 25th AIAA Aerodynamic Measurement Technology and Ground Testing Conf., AIAA-2006-3307 (2006).

Robust Estimation for Automatic Controller Tuning with Application to Active Noise Control

Charles E. Kinney and Raymond A. de Callafon

Abstract In this work, we show how the double-Youla parameterization can be used to recast the robust tuning of a feedback controller as a robust estimation problem. The formulation as an estimation problem allows tuning of the controller in real-time on the basis of closed loop data. Furthermore, robust estimation is obtained by constraining the parameter estimates so that feedback stability will be maintained during controller tuning in the presence of plant uncertainty. The combination of real-time tuning and guaranteed stability robustness opens the possibility to perform Robust Estimation for Automatic Controller Tuning (REACT) to slowly varying disturbance spectra. The procedure is illustrated via the application of narrow-band disturbance rejection in the active noise control of cooling fans.

1 Introduction

In the 60's and 70's there was a significant amount of work [8, 12, 14, 15] towards rejecting disturbances that satisfy differential equations. An interesting instance of this theory is when the differential equations have constant coefficients and poles on the imaginary axis. In this context, these results (known as the internal model principle) dictate what is required of the system so that a controller exists to reject disturbances that satisfy a known differential equation.

Today, the same principles are studied in discrete time repetitive and learning control literature [9]. The same constraints upon the system are needed as well as knowledge of the disturbance model. In practice, it is very difficult to precisely model the disturbance frequency and therefore many methods were developed to design controllers that were robust against this uncertainty in the disturbance model or to design controllers to adapt to the disturbance. Hillerström [13] used adaptive

Charles E. Kinney · Raymond A. de Callafon
Department of Mechanical and Aerospace Engineering, University of California, San Diego 9500 Gilman Drive, La Jolla, CA 92093-0411, U.S.A. e-mail: cekinney@ucsd.edu & callafon@ucsd.edu

P.M.J. Van den Hof et al. (eds.), *Model-Based Control: Bridging Rigorous Theory and Advanced Technology*, DOI: 10.1007/978-1-4419-0895-7_7,
© Springer Science + Business Media, LLC 2009

repetitive control to suppress vibrations. Bodson [3] showed an equivalence between time-varying internal models and adaptive feedforward control. Brown and Zhang [5] update the internal model to cancel a disturbance with an unknown frequency. In [4] the adaptive internal model principle is discussed. In [18], we show how the extended Kalman filter can be used to update parameters in a controller that satisfies the internal model principle. In [17], we find a family of controllers off-line that satisfy the internal model principle and a frequency estimator is used to switch between controllers online.

Landau et al. [20] used the Youla parameterization of all stabilizing controllers for a SISO system to update the controller online to reject the disturbance when the disturbance model was not completely known. It was assumed that the plant model was known exactly and the disturbances had poles on the unit circle. In [19], we added to this result by showing how to consider uncertainty to develop robust algorithms for updating the controller of SISO stable systems. The convergence of the algorithm was not analyzed.

In this work, we add to the work of [20] and [19] by considering a more general setting and proving convergence of the tuning algorithm. In this work, the plant is a MIMO system with uncertainty. The natural representation to express the uncertainty turns out to be the double-Youla parameterization [11]. This parameterization was studied by Schrama [24] in the context of cautious controller enhancement, where a controller perturbation is found off-line to improve nominal and robust performances. This parameterization is beneficial because it provides a clear strategy to maintain stability and gives access to many closed loop signals that aid in the control design. Our goal is to tune the controller in realtime to reduce the effect of the disturbances on the output and, if possible, achieve complete regulation. Attenuation of the disturbances will be accomplished in the presence of uncertainty in the plant and with very limited knowledge of the disturbance class. The only knowledge of the disturbance that will be used is the order of the generating system. The convergence of the realtime control algorithm will be analyzed be using Lyapunov theory and concepts from slowly-varying system [16]. The stability of the closed loop system will be analyzed via the Small Gain Theorem [26]. To demonstrate the effectiveness of the proposed Robust Estimation for Automatic Controller Tuning (REACT) algorithm, an experimental study based upon active noise control has been included.

2 Approach to Automatic Controller Tuning

2.1 Simultaneous Perturbation of Plant and Controller

We are considering the problem shown in Fig. 1. The output of the plant $y \in \mathbb{R}^{n_y}$, the reference signal $r \in \mathbb{R}^{n_r}$, the output of the controller $y_c \in \mathbb{R}^{n_u}$, the input disturbance $d_i \in \mathbb{R}^{n_u}$, and the output disturbance $d_o \in \mathbb{R}^{n_y}$.

The nominal plant model $G_x(s) \in \mathcal{R}_p(s)$ is a $n_y \times n_u$ transfer function matrix, where $\mathcal{R}_p(s)$ denotes the set of rational proper transfer function matrices [26], that admits a bicoprime factorization [26] given by $G_x = N_x D_x^{-1} = \tilde{D}_x^{-1} \tilde{N}_x$. The nominal controller $C(s) \in \mathcal{R}_p(s)$ is a $n_u \times n_y$ transfer function matrix that the bicoprime factorization $C = N_C D_C^{-1} = \tilde{D}_C^{-1} \tilde{N}_C$.

The uncertain system $G_\Delta \in \mathcal{R}_p(s)$ is a $n_y \times n_u$ transfer function matrix that has a dual-Youla parameterization given by

$$\begin{aligned} G_\Delta &= N_{G_\Delta} D_{G_\Delta}^{-1} = (N_x + D_c \Delta_G)(D_x - N_c \Delta_G)^{-1} \\ &= \tilde{D}_{G_\Delta}^{-1} \tilde{N}_{G_\Delta} = (\tilde{D}_x - \tilde{A}_G \tilde{N}_c)^{-1}(\tilde{N}_x + \tilde{A}_G \tilde{D}_c) , \end{aligned} \quad (1)$$

where N_{G_Δ} and D_{G_Δ} are right coprime factors $\tilde{N}_{G_\Delta}$ and $\tilde{D}_{G_\Delta}$ are the left coprime factors of the uncertain system.

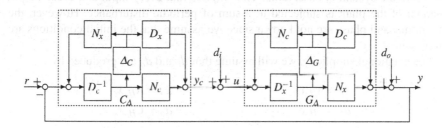

Fig. 1 Double-Youla parameterization of a feedback system with a MIMO controller C_Δ and a MIMO uncertain plant G_Δ. The controller C_Δ is expressed with the Youla parameterization and the plant G_Δ is expressed with the dual-Youla parameterization.

The uncertainty for the plant is considered to belong to the following set

$$\Xi = \{\Delta : \|\Delta\|_\infty < 1/\gamma, \ \Delta \in \mathcal{RH}_\infty\} , \quad (2)$$

for a given $\infty > \gamma > 0$. This class of uncertainty creates a set of plants Π, given by

$$\Pi = \{P : P = (N_x + D_c \Delta_G)(D_x - N_c \Delta_G)^{-1}, \Delta_G \in \Xi\} \quad (3)$$

for which a robustly stabilizing controller is sought.

The perturbed or tuned controller is described with the Youla parameterization given by

$$\begin{aligned} C_\Delta &= N_{C_\Delta} D_{C_\Delta}^{-1} = (N_C + D_x \Delta_C)(D_C - N_x \Delta_C)^{-1} \\ &= \tilde{D}_{C_\Delta}^{-1} \tilde{N}_{C_\Delta} = (\tilde{D}_C - \tilde{A}_C \tilde{N}_x)^{-1}(\tilde{N}_C + \tilde{A}_C \tilde{D}_x) , \end{aligned} \quad (4)$$

where N_{C_Δ} and D_{C_Δ} a right coprime factors of the tuned controller and $\tilde{N}_{C_\Delta}$ and $\tilde{D}_{C_\Delta}$ are the left coprime factors. The controller perturbation Δ_C is used to improve performance while maintaining robust stability in the presence of Δ_G. The perturbation

$\Delta_C \in \mathcal{RH}_\infty$ is a stable system, such that $(I - D_C^{-1}N_x\Delta_C)(\infty)$ and $(I - \tilde{\Delta}_C\tilde{N}_x\tilde{D}_C^{-1})(\infty)$ are invertible.

2.2 Disturbance Model

Each element of the input disturbance d_i and the output disturbance d_o is assumed to satisfy

$$\dot{x} = \text{diag}\left(\begin{bmatrix} 0 & \omega_1 \\ -\omega_1 & 0 \end{bmatrix}, \begin{bmatrix} 0 & \omega_2 \\ -\omega_2 & 0 \end{bmatrix}, \dots, \begin{bmatrix} 0 & \omega_{n_d} \\ -\omega_{n_d} & 0 \end{bmatrix}\right) x = A_d x, \quad x(0) = x_o$$

$$y_d = C_d x,$$

such that the system is observable. This implies that every input and every output channel of the plant is subjected to a sum of periodic disturbance. However, the magnitude and phase are not known since we assume that the initial conditions are unknown.

For notational simplicity, we will assume that d_i and d_o are produced by

$$\dot{x}_i = A_{H_i}x_i \qquad\qquad \dot{x}_o = A_{H_o}x_o$$
$$d_i = C_{H_i}x_i \qquad\qquad d_o = C_{H_o}x_o,$$

where

$$A_{H_i} = \text{diag}(\underbrace{A_d, A_d, \dots, A_d}_{n_u \text{ times}}) \quad A_{H_o} = \text{diag}(\underbrace{A_d, A_d, \dots, A_d}_{n_y \text{ times}})$$

$$C_{H_i} = \text{diag}(\underbrace{C_d, C_d, \dots, C_d}_{n_u \text{ times}}) \quad C_{H_o} = \text{diag}(\underbrace{C_d, C_d, \dots, C_d}_{n_y \text{ times}}).$$

When $\omega_i, i = 1, \dots, n_d$ are known constants, this problem can be solved with the servocompensator theory developed by Davison [7, 8], Johnson [15], Francis and Wonham [12], and more recently de Roover et al. [9]. However, the precise knowledge of the frequencies is sometimes not feasible. Therefore, in this work we will create a realtime control algorithm to reject the disturbances without knowledge of the frequencies $\omega_i, i = 1, \dots, n_d$ and maintain robust stability.

2.3 Overview of REACT

The aim of the REACT algorithm is to update the controller without violating the robustness constraint imposed by the plant uncertainty Δ_G and without complete knowledge of the disturbance model. To facilitate the automatic or realtime controller tuning, the double-Youla parameterization is used to create a robust estimation problem to find the controller perturbation Δ_C.

The REACT algorithm inputs signals from the closed loop system and outputs the controller perturbation Δ_C by using information regarding the nominal plant model and nominal controller to formulate an estimation problem. In the end, REACT provides a method for updating Δ_C to improve performance and maintain robust stability. The derivation of REACT is presented in Sec. 3, where we define and minimize a cost function. The stability and convergence of the algorithm are considered in Sec. 4. Finally, in Sec. 5 the REACT algorithm is applied to reduce the periodic noise emitted by a pair of cooling fans.

3 REACT Algorithm

3.1 Defining an Error Function

In this section, we will define an error function based upon the how the disturbances d_o and d_i effect the output y. The equation that relates the disturbances and output is given by

$$
\begin{aligned}
y &= (I + G_\Delta C_\Delta)^{-1} d_o + G_\Delta (I + C_\Delta G_\Delta)^{-1} d_i \\
&= (I + G_\Delta C_\Delta)^{-1} d_o + (I + G_\Delta C_\Delta)^{-1} G_\Delta d_i \\
&= D_{C\Delta} \tilde{\Lambda}^{-1} \tilde{D}_{G_\Delta} (d_o + G_\Delta d_i) ,
\end{aligned}
$$

where $\Lambda := \tilde{D}_{C_\Delta} D_{G_\Delta} + \tilde{N}_{C_\Delta} N_{G_\Delta}$ and $\tilde{\Lambda} := \tilde{D}_{G_\Delta} D_{C_\Delta} + \tilde{N}_{G_\Delta} N_{C_\Delta}$. After some tedious algebra we can write Λ and $\tilde{\Lambda}$ as $\Lambda = \Lambda_0 + \tilde{\Delta}_C \Lambda_0 \Delta_G$ and $\tilde{\Lambda} = \tilde{\Lambda}_0 + \tilde{\Delta}_G \Lambda_0 \Delta_C$, where $\Lambda_0 = \tilde{D}_C D_x + \tilde{N}_C N_x$ and $\tilde{\Lambda}_0 = \tilde{D}_x D_C + \tilde{N}_x N_C$.

From [23, Lemma 3.2], we get $\tilde{\Lambda} = \tilde{\Lambda}_0 + \tilde{\Lambda}_0 \Delta_G \Delta_C = \tilde{\Lambda}_0 (I + \Delta_G \Delta_C) = (I + \tilde{\Delta}_G \tilde{\Delta}_C) \tilde{\Lambda}_0$ therefore it is easy to see that we require the feedback connection of Δ_C and Δ_G to be stable if it is desired that the feedback system of G_Δ and C_Δ is stable.

Next, notice that

$$
\begin{aligned}
\tilde{\Lambda}_o^{-1} (\tilde{D}_x y - \tilde{N}_x y_c) &= \tilde{\Lambda}_o^{-1} (\tilde{D}_x D_{C\Delta} + \tilde{N}_x N_{C\Delta}) D_{C\Delta}^{-1} y = D_{C\Delta}^{-1} y \\
&= \tilde{\Lambda}^{-1} \tilde{D}_{G_\Delta} (d_o + G_\Delta d_i) ,
\end{aligned}
$$

which is obtained by substituting y and rearranging. Thus the output satisfies $y = D_{C\Delta} \tilde{\Lambda}_o^{-1} (\tilde{D}_x y - \tilde{N}_x y_c)$, where $\tilde{\Lambda}_o^{-1}$ is stable since the nominal feedback system of C and G is stable, D_x and N_x are stable, and the signals y and y_c are known.

Let η be defined as $\eta := \tilde{\Lambda}_o^{-1} (\tilde{D}_x y - \tilde{N}_x y_c)$ then the output is simply $y = (D_c - N_x \Delta_C) \eta$.

At this point, we will introduce two parameters: θ and ψ. This will facilitate in the analysis of the control algorithm by separating the optimization problem from the feedback system. Let $\Delta_C(\theta)$ indicate the parameter that will be used in the optimization problem and let $\Delta_C(\psi)$ indicate the controller perturbation that is implemented in the feedback system. In this case, we have the following error signal ε, defined as

$$\varepsilon(t, \theta, \psi) := (D_c(t) - N_x(t) * \Delta_C(t, \theta)) * \eta(t, \psi) \tag{5}$$

$$\eta(t, \psi) = \Lambda_o^{-1}(t) * (\tilde{D}_x(t) * y(t) - \tilde{N}_x(t) * y_c(t, \psi)), \tag{6}$$

where $*$ is the convolution operator.

The separation between the optimization and the feedback system is needed for convergence of the adaptive algorithm. The separation will be used to create two different time-scales, one time-scale for the feedback system and one for the optimization algorithm. The separation of time-scales is a common theme in the robustness analysis of adaptive systems [2, 22] and the same idea will be used here to prove convergence in the presence of uncertainty.

Notice that $\varepsilon(t, \psi, \psi) = y(t, \psi)$ with $\Delta_C(\psi)$ implemented. Thus, this error signal has fixed points in common with $y(t)$ but is not always equal. Also, $\varepsilon(t, \psi, \psi)$ is a nonlinear function of $\Delta_C(\psi)$ and it depends upon the uncertainty Δ_G. For these reasons it is not possible to use $\varepsilon(t, \psi, \psi)$ as an error signal, but we can use $\varepsilon(t, \theta, \psi)$ and minimize over θ.

3.2 Derivation of Algorithm

Since Δ_C appears affinely in Eq. (5), if we pick an affine parameterization of $\Delta_C(\theta)$ then $\varepsilon(t, \theta, \psi)$ will be an affine function of θ. This will aid in the analysis of robust stability when the small gain theorem is applied.

The controller perturbation has a parameterization given by

$$\Delta_C[u(t)] = \Theta(t)^T \int_0^t (C_D e^{A_D(t-\tau)} B_D + D_D) u(\tau) d\tau$$

$$= \Theta(t)^T (\mathscr{D}(t) * u(t)), \tag{7}$$

where $\Theta \in \mathbb{R}^{n_y n_\theta \times n_u}$, $\mathscr{D}$ is the stable LTI system given by

$$\mathscr{D}^T = \left[\mathbf{I}\left(\frac{s-p_0}{s+p_0}\right) \quad \mathbf{I}\left(\frac{s-p_0}{s+p_0}\right)^2 \quad \cdots \quad \mathbf{I}\left(\frac{s-p_0}{s+p_0}\right)^{n_\theta} \right],$$

and $p_0 > 0$. The state space realizations for $\mathscr{D}$ and Δ_C are given by

$$\mathscr{D} : \left[\begin{array}{c|c} A_D & B_D \\ \hline C_D & 0 \end{array}\right] \qquad \Delta_C : \left[\begin{array}{c|c} A_D & B_D \\ \hline \Theta^T(t)C_D & 0 \end{array}\right]$$

with zero initial conditions. Notice that since Δ_C is strictly proper, this parameterization guarantees that $(I - D_C^{-1} N_x \Delta_C)(\infty)$ is invertible.

If the Θ is stacked into a vector instead of a matrix then define θ as $\theta^T :=$ $\text{vec}(\Theta)^T = \begin{bmatrix} \theta_1^T & \cdots & \theta_{n_\theta}^T \end{bmatrix}$, where $\text{vec}(\cdot)$ is the vectorization operator and θ_i is the ith column of Θ. With this notation we get

$$\varepsilon(t, \Theta, \psi) = \left(D_c(t) - N_x(t) * \left(\Theta^T(t) \mathscr{D}(t) \right) \right) * \eta(t, \psi) \tag{8}$$

$$= v(t, \psi) - N_x(t) * \left(\Theta^T(t) \eta_D(t, \psi) \right), \tag{9}$$

where $v(t, \psi) := D_c(t) * \eta(t, \psi)$ and $\eta_D(t, \psi) := \mathscr{D}(t) * \eta(t, \psi)$.

Suppose that we want to minimize $V(t, \theta, \psi) = \frac{1}{2} \| \varepsilon(t, \theta, \psi) \|_2^2$ then ∇V is given by

$$\frac{\partial V(t, \theta, \psi)}{\partial \theta} = - \begin{bmatrix} [N_x]_{11} * \eta_D & \cdots & [N_x]_{n_y 1} * \eta_D \\ \vdots & \ddots & \vdots \\ [N_x]_{1 n_u} * \eta_D & \cdots & [N_x]_{n_y n_u} * \eta_D \end{bmatrix} \varepsilon(t, \theta, \psi) \tag{10}$$

$$= -\Phi(t, \psi) \varepsilon(t, \theta, \psi). \tag{11}$$

Thus, update equation is given by

$$\frac{d\theta}{dt} = \mu \Phi(t, \psi) \varepsilon(t, \theta, \psi)$$

$$= \mu r(t, \psi) - \mu \Phi(t, \psi) \Phi(t, \psi)^T \theta,$$

where $r(t, \psi) := \Phi(t, \psi) v(t, \psi)$. This is the LMS algorithm for updating θ when ψ is constant. Thus, if ψ is constrained to vary slowly then we expect similar properties to the standard LMS algorithm [2, 22].

4 Stability and Convergence of the Tuning Algorithm

In this section the stability and convergence of the realtime tuning algorithm is considered. The stability of the feedback system during tuning will be analyzed via the Small Gain Theorem [25, 26]. The convergence of the algorithm will be analyzed with Lyapunov theory [16].

4.1 Stability of the Feedback System

In this section, we investigate two different scenarios. The first is where the controller is updated very quickly. In this case, a Small Gain Theorem for time-varying systems can be used. In the second scenario, we will constrain ψ to vary slowly. In this case, we will be able to impose an LTI Small Gain Theorem for stability.

Suppose that $\Psi(t)$ is a time-varying function. If the magnitude of $\Psi(t)$ is constrained to be small enough then stability of the feedback system is assured. The following theorem clarifies this point.

Theorem 1. *Consider the feedback system depicted in Fig. 1 where the uncertain plant $G_\Delta \in \Pi$ has a representation given by Eq. (1) and the controller has a Youla-parameterization given by Eq. (4) with Δ_C given by Eq. (7) with Θ replaced by Ψ. If the time-varying matrix $\Psi(t)$ satisfies*

$$\|\Psi(t)\|_2 \leq \frac{\gamma}{\|\mathscr{D}\|_\infty} \quad \forall t$$

then the feedback system of G_Δ and C_Δ is L_2-stable for all $G_\Delta \in \Pi$.

Proof. The feedback system of G_Δ and C_Δ is stable iff the feedback system of Δ_G and Δ_c is stable since $\tilde{\Lambda}^{-1} = (I + \Delta_G\Delta_C)^{-1}\tilde{\Lambda}_0^{-1}$ and $\tilde{\Lambda}_0^{-1}$ is stable since the feedback system of G_x and C is stable. Let $u_D(t) := \mathscr{D}(t) * u(t)$ then

$$\|\Delta_C[u]\|_{L2}^2 = \int_0^\infty \|\Delta_C[u]\|_2^2 dt = \int_0^\infty \|\Psi(t)^T u_D(t)\|_2^2 dt$$

$$\leq \int_0^\infty \|\Psi(t)^T\|_2^2 \|u_D(t)\|_2^2 dt .$$

Since $\|u_D\|_{L2} \leq \|\mathscr{D}\|_\infty \|u\|_{L2}$ holds, if $\|\Psi\|_2 \leq \frac{\gamma}{\|\mathscr{D}\|_\infty}$ then

$$\|\Delta_C[u]\|_{L2}^2 \leq \frac{\gamma^2}{\|\mathscr{D}\|_\infty^2} \int_0^\infty \|u_D(t)\|_2^2 dt = \frac{\gamma^2}{\|\mathscr{D}\|_\infty^2} \|u_D(t)\|_{L2}^2$$

$$\leq \frac{\gamma^2}{\|\mathscr{D}\|_\infty^2} \|\mathscr{D}\|_\infty^2 \|u\|_{L2}^2 = \gamma^2 \|u\|_{L2}^2$$

implies that the $L2/L2$ gain of Δ_C is not greater than γ. Since $\|\Delta_G\|_\infty < 1/\gamma$ the closed loop system of Δ_G and Δ_C is L_2-stable [25]. $\square$

The result in Theorem 1 is similar to [11, Proposition 1], except here we are considering the case where the controller perturbation is a time-varying operator.

The preceding work constrains the magnitude of $\Psi(t)$ only. If a less restrictive bound is sought then additionally constraining $\dot{\Psi}(t)$ is an option. Under the right conditions, if $\dot{\Psi}(t)$ is small enough then the stability constraint is equivalent to the LTI case. Before stating the result, a lemma from obtained from [10] will be presented.

Lemma 1. *Consider $\dot{x} = A(t)x$ where $A(t)$ is a piecewise continuous function on $\mathbb{R}^+$. Suppose that $\sup_{t \geq 0} \|A(t)\| < \infty$ and $\sup_{t \geq 0} \mathscr{R}e(\text{eig}(A(t))) < 0$ then there exists an $\varepsilon > 0$ such that $\dot{x} = A(t)x$ is stable if $\sup_{t \geq 0} \|\dot{A}(t)\| \leq \varepsilon$.*

Theorem 2. *Suppose that $\|\Delta_G\|_\infty < 1/\gamma$, the frozen time system Δ_C satisfies $\|\Delta_C(\Psi)\|_\infty \leq \gamma$ for each $t \geq 0$, and that $\sup_{t \geq 0} \|\Psi(t)\| < \infty$. Then there exits an $\varepsilon > 0$ such that the feedback system of G_Δ and C_Δ is stable if $\sup_{t \geq 0} \|\dot{\Psi}(t)\| \leq \varepsilon$.*

Proof (Outline) The proof shows that the conditions for Lemma 1 are upheld. $\sup_{t \geq 0} \|\Psi(t)\| < \infty$ implies that $\sup_{t \geq 0} \|A_{cl}(t)\| < \infty$ is true, where A_{cl} is the "A" matrix of the closed loop system. $\sup_{t \geq 0} \|\Delta_G\|_\infty \|\Delta_C(\Psi)\|_\infty \leq \|\Delta_G\|_\infty \gamma < 1$ implies that $\sup_{t \geq 0} \mathcal{R}e(\text{eig}(A(t))) < 0$.

By Lemma 1 there exists an ε^* such that $\sup_{t \geq 0} \|\dot{A}_{cl}(t)\| \leq \varepsilon^*$ implies that the closed loop is stable. Due to the affine structure of the controller parameterization, for each ε^* there exits an $\varepsilon > 0$ such that $\sup_{t \geq 0} \|\dot{\Psi}(t)\| \leq \varepsilon$ implies $\sup_{t \geq 0} \|\dot{A}_{cl}(t)\| \leq \varepsilon^*$. $\square$

4.2 Convergence of the Tuning Algorithm

In this section, we analyze the convergence of the tuning algorithm by separating the time-scale of the feedback system and the update algorithm. So far, we have defined how the parameters in the optimization are updated but have not defined how the controller is updated. First it is necessary to constrain the controller perturbation parameters to be in the set of parameters that stabilize the feedback system. This can be done via the following algorithm.

$$\dot{\theta} = \mu r(t, \psi) - \mu \Phi(t, \psi) \Phi(t, \psi)^T \theta$$
$$\dot{\psi} = \underset{\psi \in \mathscr{S}}{\text{Proj}} \left(-\lambda \frac{\psi - \theta}{1 + \|\psi - \theta\|} \right),$$

where $\text{Proj}_{\psi \in \mathscr{S}}(\cdot)$ is the projection operator [2, 22] that guarantees that ψ never leaves the set of stabilizing parameters $\mathscr{S}$. The set $\mathscr{S}$ can either be

$$\mathscr{S} = \{\psi : \|\Psi\|_2 \leq \gamma / \|D\|_\infty\} \tag{12}$$

in agreement with Theorem 1 or

$$\mathscr{S} = \{\psi : \|\Delta_C(\Psi)\|_\infty \leq \gamma\} \bigcap \{\psi^T \psi \leq N_\psi < \infty\} \tag{13}$$

in agreement with Theorem 2, where N_ψ is a large number that guarantees that $\sup_{t \geq 0} \|\Psi(t)\| < \infty$. In the latter case, λ must be chosen small enough to satisfy Theorem 2. In either case, the set $\mathscr{S}$ is a compact set. Also, note that the approximation $\|\Psi\|_2 \leq \|\Psi\|_F = \|\psi\|_2$ can be used for computational speed.

To analyze this system consider for the moment ψ as being a fixed parameter and recall the error signal

$$\varepsilon(s, \theta, \psi) = (D_c - N_x \Delta_C(\theta)) \eta(s, \psi)$$
$$\eta(s, \psi) = \Lambda_o^{-1} (\tilde{D}_x y(s) - \tilde{N}_{xy} c(s, \psi))$$
$$= (D_{C\Delta}(\psi) - G_\Delta N_{C\Delta}(\psi))^{-1} (d_o + G_\Delta d_i).$$

The purpose the the tuning algorithm is to place blocking zeros [26] in the transfer function from the disturbances to the output. Thus, it is required that there exists a θ^* such that

$$D_C(j\omega_i) - N_x(j\omega_i)\Delta_C(j\omega_i, \theta^*) = \bar{D}_C(j\omega_i) = 0 \tag{14}$$

for some $\bar{D}_C$ that internally stabilizes the system and has blocking zeros at the given ω_i. From this equation it can be seen that it is required that $N_x(j\omega_i)$ have full row rank. This agrees with general servocompensator theory [15, 8, 12, 6, 9] since $N_x(j\omega_i)$ having full row rank is the same as G_x having no zeros on the $j\omega$-axis located at ω_i and at least as many inputs as outputs.

In this case, one such controller perturbation is given by

$$\Delta_C(j\omega_i, \theta^*) = N_x(j\omega_i)^T (N_x(j\omega_i)N_x(j\omega_i)^T)^{-1}D_C(j\omega_i, \theta^*) .$$

When $\Delta_C = \Theta^T \mathscr{D}$ then Θ given by

$$\Theta^T = N_x(j\omega_i)^+ D_C(j\omega_i, \theta^*)\mathscr{D}(j\omega_i)^+$$

will satisfy Eq. (14), where it is required that $\mathscr{D}(j\omega_i) \in \mathbb{C}^{n_\theta n_y \times n_y}$ has full column rank, $N_x(j\omega_i)^+ = N_x(j\omega_i)^T (N_x(j\omega_i)N_x(j\omega_i)^T)^{-1}$, and $\mathscr{D}(j\omega_i)^+ = (\mathscr{D}(j\omega_i)^T \mathscr{D}(j\omega_i))^{-1}\mathscr{D}(j\omega_i)^T$.

Theorem 3. *Consider the controller perturbation Δ_C given by Eq. (7). Suppose that $N_x(j\omega_i)$ has full row rank for all $0 \leq i \leq n_d$ and that $n_\theta \geq 2n_d$.*
Define the following matrices: $D := \begin{bmatrix} \mathscr{D}(j\omega_1) & \mathscr{D}(j\omega_2) & \cdots & \mathscr{D}(j\omega_{n_d}) \end{bmatrix}$,

$$b^T := \begin{bmatrix} D_C(j\omega_1)^T (N_x(j\omega_1)^+)^T \\ D_C(j\omega_2)^T (N_x(j\omega_2)^+)^T \\ \vdots \\ D_C(j\omega_{n_d})^T (N_x(j\omega_{n_d})^+)^T \end{bmatrix}, \quad A := \begin{bmatrix} \mathscr{R}\mathrm{e}(D^T) \\ \mathrm{im}(D^T) \end{bmatrix}, \quad and \quad B := \begin{bmatrix} \mathscr{R}\mathrm{e}(b^T) \\ \mathrm{im}(b^T) \end{bmatrix} .$$

Then A is full rank if $\omega_i \neq \omega_j$ for all $i \neq j$ and in this case the optimal parameter given by $\Theta^* = A^+ B$ satisfies $D_C(j\omega_i) - N_x(j\omega_i)(\Theta^*)^T \mathscr{D}(j\omega_i) = 0$, for all $0 \leq i \leq n_d$.

Proof. For the controller perturbation given in Eq. (7) D can be written as

$$D = I_{n_y \times n_y} \otimes \begin{bmatrix} \frac{j\omega_1-p_o}{j\omega_1+p_o} & \frac{j\omega_2-p_o}{j\omega_2+p_o} & \cdots & \frac{j\omega_{n_d}-p_o}{j\omega_{n_d}+p_o} \\ \left(\frac{j\omega_1-p_o}{j\omega_1+p_o}\right)^2 & \left(\frac{j\omega_2-p_o}{j\omega_2+p_o}\right)^2 & \ddots & \left(\frac{j\omega_{n_d}-p_o}{j\omega_{n_d}+p_o}\right)^2 \\ \vdots & \vdots & \ddots & \vdots \\ \left(\frac{j\omega_1-p_o}{j\omega_1+p_o}\right)^{n_\theta} & \left(\frac{j\omega_2-p_o}{j\omega_2+p_o}\right)^{n_\theta} & \cdots & \left(\frac{j\omega_{n_d}-p_o}{j\omega_{n_d}+p_o}\right)^{n_\theta} \end{bmatrix},$$

where $\otimes$ is the Kronecker product.

Thus, $[D\ \bar{D}]$ is the first $n_d n_y$ columns of a block Vandermonde matrix which can be written as $[D\ \bar{D}] = I \otimes d$ where d is the first n_d columns of a Vandermonde matrix and $(\bar{\cdot})$ is the complex conjugate. The Vandermonde matrix is invertible if $\omega_i \neq \omega_j$ for all $i \neq j$. Thus $[D\ \bar{D}]$ is full rank. And since

$$[D \; \bar{D}] \begin{bmatrix} \mathbf{I}\frac{1}{2} & \mathbf{I}\frac{1}{2j} \\ -\mathbf{I}\frac{1}{2} & -\mathbf{I}\frac{1}{2j} \end{bmatrix} = [\mathscr{R}e(D) \; im(D)]$$

then $[\mathscr{R}e(D) \; im(D)]$ is full rank.

The equation $(\Theta^*)^T[\mathscr{R}e(D) \; im(D)] - [\mathscr{R}e(b) \; im(b)] = 0$ can be written as

$$(\Theta^*)^T[D \; \bar{D}] \begin{bmatrix} \mathbf{I}\frac{1}{2} & \mathbf{I}\frac{1}{2j} \\ -\mathbf{I}\frac{1}{2} & -\mathbf{I}\frac{1}{2j} \end{bmatrix} = [b \; \bar{b}] \begin{bmatrix} \mathbf{I}\frac{1}{2} & \mathbf{I}\frac{1}{2j} \\ -\mathbf{I}\frac{1}{2} & -\mathbf{I}\frac{1}{2j} \end{bmatrix} ,$$

and since $\begin{bmatrix} \mathbf{I}\frac{1}{2} & \mathbf{I}\frac{1}{2j} \\ -\mathbf{I}\frac{1}{2} & -\mathbf{I}\frac{1}{2j} \end{bmatrix}$ is invertible then we get $(\Theta^*)^T[D \; \bar{D}] = [b \; \bar{b}]$. From the

first part of this matrix, Θ^* satisfies $(\Theta^*)^T D = b$ which is the same as

$$(\Theta^*)^T = N_x(j\omega_i)^+ D_C(j\omega_i, \theta^*)\mathscr{D}(j\omega_i)^+ ,$$

for all $0 \le i \le n_d$ since $N_x(j\omega_i)$ has full row rank and $\mathscr{D}(j\omega_i)$ has full column rank when $\omega_i \ne \omega_j$ for all $i \ne j$. This implies that

$$D_C(j\omega_i) \quad N_x(j\omega_i)(\Theta^*)^T \mathscr{D}(j\omega_i) = 0 ,$$

for all $0 \le i \le n_d$. $\square$

In the case that the ω_i's are known then this theorem provides an equation for θ^* that can be implemented. In this work, we are considering the ω_i's as unknown and are using the realtime tuning algorithm to converge to this point θ^*. Using this optimal point θ^*, the error signal is written as

$$\varepsilon(s, \theta, \psi) = (D_c - N_x \Delta_C(\theta))\eta(s, \psi)$$
$$= (\bar{D}_C(s) + N_x(\Delta_C(\theta^*) - \Delta_C(\theta)))\eta(s, \psi) ,$$

where $\bar{D}_C$ is given in Eq. (14). In the time domain, this can be written as

$$\varepsilon(t, \theta, \psi) = \bar{D}_C(t) * \eta(t, \psi) + N_x(t) * \left((\Theta^* - \Theta(t))^T \eta_D(t, \psi)\right) ,$$

where $\bar{D}_C(t) * \eta(t, \psi) \to 0$ since $\bar{D}_C(s) \in \mathscr{R}\mathscr{H}_\infty$ and has blocking zeros at ω_i.

Hence, the point θ^* such that $D_c(j\omega_i) - N_x(j\omega_i)\Delta_C(j\omega_i, \theta^*) = 0$ is an equilibrium point for all fixed $\psi \in \Gamma$. Next, change variables $x = \theta - \theta^*$ to get

$$\dot{x} = \mu r(t, \psi) - \mu \Phi(t, \psi)\Phi(t, \psi)^T(x + \theta^*)$$
$$= -\mu \Phi(t, \psi)\Phi(t, \psi)^T x + v(t, \psi) ,$$

where $v(t, \psi)$ is an exponentially decreasing signal due to initial conditions. The solution $x \to 0$ if $v(t, \psi) \to 0$ exp. fast and $\dot{x} = -\mu \Phi(t, \psi)\Phi(t, \psi)^T x$ is exp. stable. So, by studying the stability of

$$\dot{z} = -\mu \Phi(t, \psi) \Phi(t, \psi)^T z$$

we can determine when the optimization will converge to the desired solution.

To analyze the convergence of the algorithm, we will find a Lyapunov function. To this end we first start by examining when a suitable Lyapunov function exists for a given system.

Lemma 2. *Let* $x = 0$ *be an equilibrium for*

$$\dot{x} = f(t, x, \alpha) \, ,$$

where $f : [0, \infty) \times D \times \gamma \rightarrow \mathbb{R}^n$ *is continuously differentiable,* $D = \{x \in \mathbb{R}^n : \|x\| < r\}$, *and*

$$\left\| \frac{\partial f}{\partial x} \right\| \le L_1 \qquad \left\| \frac{\partial f}{\partial \alpha} \right\| \le L_2 \|x\|$$

on D, *uniformly in* t *and* α. *Let* k, λ, *and* r_o *be positive constants with* $r_o < r/k$. *Let* $D_o = \{x \in \mathbb{R}^n : \|x\| < r_o\}$. *Assume that the trajectories of the system satisfy*

$$\|x(t)\| \le k \|x(t_0)\| e^{-\lambda(t - t_0)} \quad \forall x \in D, t \ge t_0 \ge 0, \alpha \in \Gamma$$

then there exists a function $V : [0, \infty) \times D_o \times \gamma \rightarrow \mathbb{R}$ *such that the following hold*

$$c_1 \|x\|^2 \le V(t, x, \alpha) \le c_2 \|x\|^2, \quad \frac{\partial V}{\partial t} + \frac{\partial V}{\partial x} f(t, x, \alpha) \le -c_3 \|x\|^2, \quad \left\| \frac{\partial V}{\partial \alpha} \right\| \le c_4 \|x\|^2 \, .$$

Proof (Outline) Let $\phi(\tau, t, x, \alpha)$ indicate the solution to $\dot{x}(t) = f(t, x(t), \alpha)$ starting at (t, x) for a given $\alpha \in \Gamma$. Thus $\phi(t, t, x, \alpha) = x$. Choose

$$V(t, x, \alpha) = \int_t^{t + \delta} \phi(\tau, t, x, \alpha)^T \phi(\tau, t, x, \alpha) d\tau$$

as the Lyapunov function. It can be shown that this Lyapunov function satisfies the inequalities, and the proof is similar to proofs of Theorem 4.14 and Lemma 9.18 in [16]. □

Now, we can state the convergence result.

Theorem 4. *Suppose that* $\Phi(t, \psi)$ *is uniformly P.E., i.e. there exists constants* β, δ, *and* T *such that*

$$\infty > \beta I \ge \int_{t_0}^{t_0 + T} \Phi(t, \psi) \Phi(t, \psi)^T dt \ge \delta I > 0$$

holds for all $t_0 \ge 0$, *all fixed* $\psi \in \mathscr{S}$, *and where* β *and* δ *are independent of* ψ. *Then there exists an* $\varepsilon^* > 0$ *such that*

$$\dot{z} = -\mu \Phi(t, \psi(t)) \Phi(t, \psi(t))^T z, \qquad \psi : \mathbb{R}^+ \rightarrow \mathscr{S}$$

converges to the origin exponentially fast if $\|\dot{\psi}\| \leq \varepsilon^*$.

Proof (Outline) Since the system is continuously differentiable and driven by a bounded periodic signal it can be shown that the assumptions of Lemma 2 are upheld. Therefore, there exists a Lyapunov function s.t.

$$c_1\|z\|^2 \leq V(z,\psi,t) \leq c_2\|z\|^2$$

$$\frac{\partial V}{\partial t} + \frac{\partial V}{\partial z}\dot{z} \leq -c_3\|z\|^2$$

$$\left\|\frac{\partial V}{\partial \psi}\right\| \leq c_4\|z\|^2$$

hold. Using this function $V(z,\psi(t))$ as the Lyapunov function for

$$\dot{z} = -\mu \Phi(t,\psi(t))\Phi(t,\psi(t))^T z,$$

where now $\psi(t)$ is a function of t instead of a fixed parameter, yields

$$\dot{V}(z,\psi,t) = \left\|\frac{\partial V}{\partial t} + \frac{\partial V}{\partial z}\dot{z} + \frac{\partial V}{\partial \psi}\dot{\psi}\right\| \leq -c_3\|z\|^2 + \left\|\frac{\partial V}{\partial \psi}\dot{\psi}\right\|$$

$$\leq -c_3\|z\|^2 + \left\|\frac{\partial V}{\partial \psi}\right\|\|\dot{\psi}\| \leq -c_3\|z\|^2 + c_4\|z\|^2\|\dot{\psi}\|$$

$$= (c_4\|\dot{\psi}\| - c_3)\|z\|^2.$$

Therefore if $\|\dot{\psi}\| < c_3/c_4 := \varepsilon^*$ then $z = 0$ is exponentially stable. $\square$

This theorem states when θ converges to the correct point θ^*. To conclude when regulation will occur the set $\mathscr{S}$ must contain θ^* and the point θ^* must exist in the first place. Thus we have the following.

Theorem 5. *Consider the feedback system depicted in Fig. 1 where the uncertain plant $G_\Delta \in \Pi$ has a representation given by Eq. (1) and the controller has a Youla-parameterization given by Eq. (4) with Θ replaced by Ψ. Assume that the controller perturbation parameters ψ are updated with the following tuning algorithm*

$$\dot{\theta} = \mu r(t,\psi) - \mu \Phi(t,\psi)\Phi(t,\psi)^T \theta$$

$$\dot{\psi} = \operatorname*{Proj}_{\psi \in \mathscr{S}}\left(-\varepsilon \frac{\psi-\theta}{1+\|\psi-\theta\|}\right),$$

where $\mathscr{S}$ is defined in either Eq. (12) or Eq. (13). If $N_x(j\omega_i)$ has full row rank for all $0 \leq i \leq n_d$, $n_\theta \geq 2n_d$, and $\Phi(t,\psi)$ is uniformly P.E. then there exists an $\varepsilon^* > 0$ such that $\lim_{t\to\infty}\|\theta(t) - \theta^*\| = 0$ whenever $\varepsilon \leq \varepsilon^*$, where θ^* is given in Theorem 3, and if $\theta^* \in \mathscr{S}$ then $\lim_{t\to\infty}\|\psi(t) - \theta^*\| = 0$ which implies $\lim_{t\to\infty}\|y(t)\| = 0$.

Proof. Straightforward combination of Theorem 3 and Theorem 4, where either Theorem 1 or Theorem 2 is used to show that the set $\mathscr{S}$ is a set of robustly stabilizing parameters. $\square$

5 Application to ANC

5.1 Description of System

Fig. 2 shows the layout of the acoustic system that we are considering. Two variable speed cooling fans connected in series are used to cool the enclosure, similar to a server or PC. However, due to the high speeds of the fans, acoustic noise is created. To combat the acoustic noise, 4 speakers are mounted near the fan and 4 feedback microphones are placed near the speakers and downstream of the acoustic noise. To reduce vibrations and turbulent noise, the microphones are mounted in acoustical foam. Additionally, for simplicity, the microphone signals are summed together and used as a single signal for feedback. Similarly, the same signal is sent to the speakers so that we are dealing with a single-input-single-output (SISO) system.

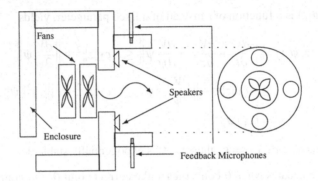

Fig. 2 Custom fan housing with active noise canceling speakers and feedback microphones.

The fans create two types of noise: broadband and narrowband noise. The narrowband noise is due to the blade pass frequency (BPF) of each fan and is comprised of a fundamental frequency and several harmonics. The broadband noise is due to turbulence. Both types of noise are dependent upon the speed of the fans. When the RPM of a cooling fan is increased the BPF increases causing the fundamental frequency to increase. Likewise, when the RPM increases, the turbulence increases and therefore the broadband noise level will increase.

The goal of the active noise control system is to reduce the narrowband acoustic noise without a priori knowledge of the fan speeds and with the presence of modeling errors. The REACT algorithm described in Sec. 2.3 is used to this end. The identification of a nominal model for the REACT algorithm is presented in Sec. 5.2 and the experimental results of applying REACT to the ANC system are presented in Sec. 5.3.

5.2 Identification of Plant Model

The nominal plant model G_x is found via standard system identification techniques [21]. To generate data that can be used for the identification, a white noise signal was sent into the speakers and the resulting signal was recorded with the feedback microphones. Recall that the microphone signals are averaged and the speakers are sent the same signal so that a SISO system results, and note that the acoustic system is comprised of the speaker amplifier, the speakers, that acoustic between the speaker and microphone, the microphone, and the microphone filter.

At steady state, if $\|\Delta_C \Delta_G\|_\infty < 1$ then the closed loop system is stable by the small gain theorem [26]. In addition, the internal model principle requires that $D_C(j\omega_k) - N_x(j\omega_k)\Delta_C(j\omega_k) = 0$, $k = 1, 2, ..., n_d$ for complete regulation, where ω_k is the frequency of the disturbances. Since we are dealing with a SISO, open-loop stable system we can choose $N_x = G_x$, $D_x = 1$, $N_C = 0$, and $D_C = 1$. Let G_o denote the "true" system then the requirements for stability and performance are given by the following:

1. $|G_o(e^{j\omega}) - G_x(e^{j\omega})| = |\Delta_{G_o}(e^{j\omega})| < 1/|\Delta_C(e^{j\omega})|$ for all ω.
2. $G_x(e^{j\omega_o}) = 1/\Delta_C(e^{j\omega_k})$ where ω_o is the unknown frequency of the disturbance.

Hence, these requirements can be satisfied if $|\Delta_{G_o}(e^{j\omega})| < |G_x(e^{j\omega})|$ holds over the frequency range that the disturbance is expected. A nominal model that satisfies this bound is deemed acceptable.

After several iterations a suitable model was chosen by the above design method. Using an ARX model structure [21] and a Steiglitz-Mcbride iteration a 25th order model of the acoustic system was found. The frequency response of the model G_x and the true system G_o is shown in Fig. 3. It can be seen that the model neglects some low and high frequency dynamics and is a close approximation of the frequency response measurements in the middle frequencies. The nominal controller is chosen as $C = 0$ since the model is open-loop stable.

The magnitude of the uncertainty $|\Delta_{G_o}|$ is also shown in Fig. 3. In this figure, it can be seen that regulation is possible from 300 to 5000 Hz. However, the Δ_C that is found must satisfy $|\Delta_{G_o}(e^{j\omega})| < 1/|\Delta_C(e^{j\omega})|$ for all ω and one must be careful not to violate this bound while tuning the controller.

5.3 Experimental Results

The REACT algorithm described in Sec. 2.3 is applied to the ANC system described in Sec. 5.1. The results of the controller tuning are shown in Fig. 4. In this figure, the A-weighted[1] output $y(t)$ and control $u(t)$ signals are shown. On the left, the signals are shown for a duration of 6.4 sec. The first 3.2 sec. are without controller tuning. The last 3.2 sec. are the signals after the convergence of REACT. On the right, the

---

[1] A-weighting is used to reflect the sensitivity of human hearing [1].

Fig. 3 Frequency response data of G_o (dotted), bode plot the nominal system G_x (solid) found with system identification techniques [21], and the magnitude of the additive uncertainty $|\Delta_{G_o}|$ (dashed). This figure is used to evaluate the quality of the nominal model G_x. For complete regulation in the presence of uncertainty, it is required that $|\Delta_{G_o}(e^{j\omega})| < |G_x(e^{j\omega})|$ for the frequency range of interest.

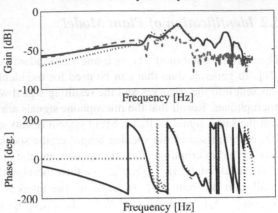

signals are shown for the first and last 0.0195 sec. In the plots on the left, it is clear that the level of the output signal is reduced when the controller tuning is switched on. The removal of the narrowband disturbances can be seen clearly in the plots on the right.

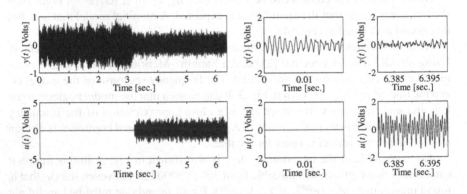

Fig. 4 The A-weighted output $y(t)$ and input $u(t)$ of the plant before and after tuning of the nominal controller. On the left is the entire 6.4 sec of data. On the right is the output and input for the first and last 0.0195 sec. The tuning algorithm reduces the output variance by 79%, or a 6.7 dB drop in SPL, by eliminating narrowband disturbances.

The A-weighted sample variance of the signal is reduced by 79%, which is a 6.7dB drop in the A-weighted sound pressure level (SPL)[2]. The A-weighted spectrum of the output $y(t)$ before and after tuning is shown in Fig. 5. In this figure, it can be seen that largest harmonics are reduced. The reduction of these harmonics

---

[2] SPL is defined as $SPL := 10\log_{10}(P^2/P_0^2)$ where P^2 is the sample variance of the microphone signal and P_0^2 is the reference level. If weighting is applied then the signal is filtered before the calculation of the sample variance.

can also be seen in Fig. 4. Note that both fans are running at similar speeds and therefore two harmonics from each fan, for a total of 4 harmonics, are reduced by tuning the controller.

Fig. 5 A-weighted spectrum of the plant output $y(t)$ (the microphone signal) before (dashed) and after (solid) controller tuning. The tuning of the controller reduces the most prominent harmonics of both cooling fans. The result is a reduction in the A-weighted output variance by 79%, or a 6.7 dB drop in A-weighted SPL.

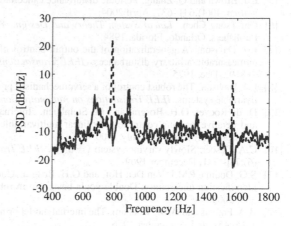

6 Conclusions

We have shown how to tune a given nominal controller in realtime to reject disturbances and to preserve robust stability. The uncertain system was assumed to belong to a set of systems described with the dual-Youla parameterization. To facilitate the tuning of the controller, the Youla parameterization was used along with a realtime optimization that utilizes signals from the closed loop system and information regarding the nominal system. The magnitude and velocity of the controller perturbation parameters are constrained to preserve robustness, and to guarantee that the optimization converges the controller in the feedback system is changed gradually. An application to active noise control was used to demonstrate the power of the proposed tuning algorithm. In the experimental results, it was shown how the harmonic noise of cooling fans can be reduced in the presence of modeling errors and measurement noise without a priori knowledge of the harmonic noise frequency.

References

[1] *Occupational Health & Safety Administration 29 CFR*. Part 1910, Standard 1910.95: Occupational Noise Exposure.
[2] B.D.O. Anderson, R.R. Bitmead, C.R. Johnson, P.V. Kokotović, R.L. Kosut, I.M.Y. Mareel, L. Praly, and B.D. Riedle. *Stability of Adaptive Systems: Passivity and Averaging Analysis*. MIT Press, Cambridge, MA, USA, 1986.
[3] M. Bodson. Equivalence between adaptive cancellation algorithms and linear time-varying compensators. In *The 43rd IEEE Conference on Decision and Control*, pages 638–643, 2004.

[4]   M. Bodson. Rejection of periodic disturbances of unknown and time-varying frequency. *International Journal of Adaptive Control and Signal Processing*, 19(2-3):67–88, March–April 2005.

[5]   L.J. Brown and Q. Zhang. Periodic disturbance cancellation with uncertain frequency. *Automatica*, 40(4):631–637, April 2004.

[6]   Chi-Tsong Chen. *Linear System Theory and Design*. Harcourt Brace Jovanovich College Publishers, Orlando, Florida, 1984.

[7]   E.J. Davison. A generalization of the output control of linear multivariable systems with unmeasurable arbitrary disturbances. *IEEE Transactions on Automatic Control*, AC-20(6): 788–92, Dec. 1975.

[8]   E.J. Davison. The robust control of a servomechanism problem for linear time-invariant multivariable systems. *IEEE Transactions on Automatic Control*, AC-21(1):25–34, Feb. 1976.

[9]   D. de Roover, O.H. Bosgra, and M. Steinbuch. Internal-model-based design of repetitive and iterative learning controllers for linear multivariable systems. *International Journal of Control*, 73(10):914–929, July 2000.

[10] C.A. Desoer. Slowly varying system $\dot{x} = a(t)x$. *IEEE Transactions on Automatic Control*, 14 (6):780–781, December 1969.

[11] S.G. Douma, P.M.J. Van Den Hof, and O.H. Bosgra. Controller tuning freedom under plant identification uncertainty: Double youla beats gap in robust stability. *Automatica*, 39:325–333, 2003.

[12] B.A. Francis and W.M. Wonham. The internal model principle of control theory. *Automatica*, 12(5):457–465, September 1976.

[13] G. Hillerström. Adaptive suppression of vibrations - a repetitive control approach. *IEEE Trans. on Control Systems Tech.*, 4(1):72–8, January 1996.

[14] C.D. Johnson. Optimal control of the linear regulator with constant disturbances. *IEEE Trans. on Automatic Control*, 13(4):416–421, August 1968.

[15] C.D. Johnson. Accommodation of external disturbances in linear regulator and servomechanism problems. *IEEE Trans. on Automatic Control*, 16(6):635–644, December 1971.

[16] H.K. Khalil. *Nonlinear Systems*. Prentice Hall, Upper Saddle River, NJ USA, 3rd edition, 2002.

[17] C.E. Kinney and R.A. de Callafon. An adaptive internal model-based controller for periodic disturbance rejection. In 14th *IFAC Symposium on System Identification*, pages 273–278, Newcastle, Australia, March 2006.

[18] C.E. Kinney and R.A. de Callafon. Adaptive periodic noise cancellation for cooling fans with varying rotational speed. In *IDETC/CIE 2007*, Las Vegas, NV, September 4th-7th 2007. DETC2007-35730.

[19] C.E. Kinney, A. Villalta, and R.A. de Callafon. Active noise control of a cooling fan in a short duct. In *Proceedings of NCAD2008 NoiseCon2008-ASME NCAD*, Dearborn, MI, 2008. NCAD2008-73086.

[20] I.D. Landau, A. Constantinescu, and D. Rey. Adaptive narrowband disturbance rejection applied to an active suspension-an internal model principle approach. *Automatica*, 41(4): 563–574, January 2005.

[21] L. Ljung. *System Identification-Theory for the User*. Prentice-Hall, Englewood Cliffs, NJ USA, second edition, 1999.

[22] S. Sastry and M. Bodson. *Adaptive Control : Stability, Convergence, and Robustness*. Prentice Hall, Englewood Cliffs, NJ, USA, 1989.

[23] R.J.P. Schrama, P.M.M. Bongers, and O.H. Bosgra. Robust stability under simultaneous perturbations of linear plant and controller. In *Proceedings of the 31st Conf. On Decision and Control*, pages 2137–2139, Tucson, AZ, 1992.

[24] Ruud Schrama. *Approximate Identification and Control Design with Application to a Mechanical System*. PhD thesis, Delft University of Tech., 1992.

[25] A. Van der Schaft. *L2-Gain and Passivity Techniques in Nonlinear Control*. Lecture Notes in Control and Information Sciences. Springer-Verlag Berlin Heidelberg, New York, 1996.

[26] K. Zhou, J.C. Doyle, and K. Glover. *Robust and Optimal Control*. Prentice-Hall, Upper Saddle River, NJ, 1996.

Identification of Parameters in Large Scale Physical Model Structures, for the Purpose of Model-Based Operations

Paul M.J. Van den Hof, Jorn F.M. Van Doren and Sippe G. Douma

Abstract When first principles models are used for model-based operations as monitoring, control and optimization, the estimation of accurate physical parameters is important in particular when the underlying dynamical model is nonlinear. If the models are the result of partial differential equations being discretized, they are often large-scale in terms of number of states and possibly also number of parameters. Estimating a large number of parameters from measurement data leads to problems of identifiability, and consequently to inaccurate identification results. The question whether a physical model structure is identifiable, is usually considered in a qualitative way, i.e. it is answered with a yes/no answer. However since also nearly unidentifiable model structures lead to poor parameter estimates, the questions is addressed how the model structure can be approximated so as to achieve local identifiability, while retaining the interpretation of the physical parameters. Appropriate attention is also given to the relevant scaling of parameters. The problem is addressed in a prediction error setting, showing the relation with gradient-type optimization algorithms as well as with Bayesian parameter estimation.

Paul M.J. Van den Hof
Delft Center for Systems and Control, Delft University of Technology, Mekelweg 2, 2628 CD
Delft, The Netherlands, e-mail: p.m.j.vandenhof@tudelft.nl

Jorn F.M. Van Doren
Delft Center for Systems and Control, Delft University of Technology, Mekelweg 2, 2628 CD
Delft, The Netherlands, e-mail: j.f.m.vandoren@tudelft.nl and Shell International E&P, Kessler
Park 1, 2288 GS Rijswijk, The Netherlands, e-mail: jorn.vandoren@shell.com

Sippe G. Douma
Shell International Exploration and Production, P.O. Box 60, 2280 AB Rijswijk, The Netherlands,
e-mail: sippe.douma@shell.com

P.M.J. Van den Hof et al. (eds.), *Model-Based Control: Bridging Rigorous Theory
and Advanced Technology*, DOI: 10.1007/978-1-4419-0895-7_8,
© Springer Science + Business Media, LLC 2009

1 Introduction

Complex dynamical physical processes raise many challenges for model-based monitoring, control and optimization. On-line reconstruction of non-measurable variables, design of appropriate feedforward and feedback control strategies, as well as economic optimization of processes under appropriate operational constraints, generally require the availability of a reliable process model, preferably accompanied by a quantification of its reliability (uncertainty). If the dynamics of the considered process is linear, then a process model can be obtained by applying black-box system identification, which provides a well-studied set of tools for identifying linear models on the basis of experimental data [15, 21]. If there is a particular interest in the identification of physical parameters, this often does not raise any additional problems: one has to choose the right (physics-based) model structure and identify the parameters through one of the available (possibly non-convex) optimization methods. The only issue that has to be taken care of is that the physical model structure is identifiable, implying that the several physical parameters can be distinguished from each other on the basis of the model's input-output behavior.

In the situation that the process dynamics is nonlinear, it can often be linearized around an operating point (as e.g. in continuous-type industrial/chemical production processes) and the above mentioned linear approach can be followed leading to a linear (approximate) model. However when essential nonlinear dynamical phenomena are involved and the user needs to capture this dynamics in the model, it is much harder to come up with generic black-box techniques for identification. Although there are interesting attempts to capture the nonlinear phenomena in (black-box) model structures as Wiener and/or Hammerstein models, [1, 33], and linear parameter-varying (LPV) models [14, 29, 2, 25, 24, 32, 34], information on the underlying physical structure of the nonlinearities is very often required for selection of an appropriate model structure.

In some processes it is desirable to capture the real underlying nonlinear dynamic structure of the process in order to make reliable long-term predictions. First-principles model then provide the structure of the model, while incorporated (physical) parameters have to be estimated on the basis of experimental data. Especially in situations where the first principles models are given by partial differential equations (pde's), the required step of discretizing the equations in space and time generally leads to complex models with a large number of states and possibly also a large number of unknown (physical) parameters. For an interesting example of this situation in a problem of (oil) reservoir engineering, the reader is referred to [13], where an industrial example of optimal oil recovery from reservoirs over the life time of the reservoir (possibly > 20 years) is shown in the form of a nonlinear (batch-type) process with a number of states and parameters exceeding the order of 10^5.

Identifying extremely large number of parameters from measurement data leads to serious problems, and at least it leads to the question which model properties can be reliably estimated from the available measurement data. From a model-based operations point of view (monitoring, control, optimization) it makes sense to limit the complexity of an identified model to a level where the model can be reliably validated from data. If not, the parameter estimates might be highly determined by the -random- experiment that is done (overfit) leading to unreliable model predictions. In identification this problem is addressed by the notion of identifiability (of a model structure), and directly coupled to the variance of estimated parameters.

In this chapter the notion of identifiability will be evaluated in the scope of the high-complexity type of processes discussed above. Our argument will be that having an (locally) identifiable model structure will not be sufficient to provide reliable parameter estimates in large scale physical model structures. Methods will be presented that allow to reduce the parameter space to limited dimension, while being able to reliably estimate the reduced parameters and maintaining their physical interpretation. To this end the qualitative notion of (local) identifiability (with a yes/no answer) is generalized to a quantitative notion, removing that parameter subspace from the parametrization that can only be estimated with excessively large variance.

2 Identifiability - the starting point

The notion of identifiability refers -roughly speaking- to the question whether parameter changes in the model can be observed in the model output signal (output identifiability) or in the model's input-output map or transfer function (structural identifiability).

The notion of output identifiability has been studied in e.g. [11] and [15]. The notion of structural identifiability was first stated by [3] and has been extensively studied in the field of compartmental modeling ([10], [20]). State-space model parameterizations have been analyzed by [9] and [30]. Lately there has been a renewed interest in structural identifiability analysis, with contributions from [22] and [27]. In its essence, identifiability properties are global properties, i.e. holding for the full parameter space, as e.g. considered in [16] and [7] where parameter mappings are studied to analyse global identifiability. However restricting attention to a local analysis is often the only situation that is feasible in terms of computational complexity. As a result we will focus on local properties in this chapter only.

Consider a nonlinear dynamical model that generates output predictions according to[1]:

$$\hat{\mathbf{y}} = h(\mathbf{u}, \theta; x_0), \tag{1}$$

where $\hat{\mathbf{y}}$ is a prediction of $\mathbf{y} := \begin{bmatrix} y_1^T & \dots & y_N^T \end{bmatrix}^T$ denoting output signal measurements $y_k \in \mathbb{R}^p$ stacked over time, $\theta \in \Theta \subset \mathbb{R}^q$ the parameter vector, $\mathbf{u} := \begin{bmatrix} u_1^T & \dots & u_N^T \end{bmatrix}^T$ the input vector $u_k \in \mathbb{R}^m$ stacked over time, and x_0 the initial state vector. Since the model (1) is parameterized it represents an input/output model structure.

The definition of local identifiability now is given as follows([11]):

Definition 1 *An input/output model structure $h(\theta, \mathbf{u}; x_0) : \Theta \to \mathscr{H}$ is called locally identifiable in $\theta_m \in \Theta$ for a given $\mathbf{u}$ and x_0, if for all θ_1, θ_2 in the neighborhood of*

---

[1] Without loss of generality we restrict attention to predictors that are not dependent on output measurements y, whch in an LTI-setting is referred to as Output Error predictors.

θ_m holds that

$$\{h(\mathbf{u}, \theta_1; x_0) = h(\mathbf{u}, \theta_2; x_0)\} \Rightarrow \theta_1 = \theta_2.$$

If we linearize the nonlinear process dynamics around a chosen operating point or trajectory, a linear dynamical system results. This system can be modelled by an LTI input-output model, represented by the transfer function G, leading to an output predictor

$$\hat{y}_k = G(q, \theta) u_k,$$

with q the shift operator $q u_k = u_{k+1}$.

The notion of local structural identifiability ([9]) is then defined by considering the properties of the parameterized transfer function $G(q, \theta)$:

Definition 2 *An input/output model structure* $G: \Theta \rightarrow \mathscr{G}$ *with* $\Theta \subset \mathbb{R}^q$ *and* $\mathscr{G} \subset \mathbb{R}(z)^{p \times m}$ *is called locally structural identifiable in* $\theta_m \in \Theta$ *if for all* θ_1, θ_2 *in the neighborhood of* θ_m *holds that*

$$\{G(z, \theta_1) = G(z, \theta_2)\} \Rightarrow \theta_1 = \theta_2. \tag{2}$$

Here $\mathbb{R}(z)^{p \times m}$ is the set of all $p \times m$ rational transfer function matrices in the complex indeterminate z.

Note that in contrast with (1) this latter notion is not dependent on either an input signal nor an initial state. Structural identifiability will be considered in section 8, where a link is made between structural identifiability and identifiability.

In general identifiability questions are considered qualitatively, i.e. deciding whether a model structure is either identifiable or not. The tests required for this evaluation are typically rank evaluations of matrices, as e.g. Fisher's information matrix, around a particular local operating point in the parameter space, see e.g. [6]. However, when considering parameters in large scale (nonlinear) physical models it is relevant to raise the question how the notion of identifiability can be quantified. This implies addressing the question which part of the parameter space is best iden- tifiable, and which part of the model structure can be approximated so as to achieve local identifiability, while retaining the interpretation of the physical parameters. For structural identifiability this question was preliminary addressed in [27]. In [4] the degree of identifiability was introduced. In [26] principal component analysis was applied to determine which parameters can be identified. Assessing identifiability can also be done a posteriori, after the identification of all parameters, by evaluating the parameter variance, see e.g. [12].

In this chapter we will further investigate how the notions of identifiability can be quantified to allow for a reduction in the parameter space with physically inter- pretable parameters.

3 Testing local identifiability in identification

3.1 Introduction

In a model identification framework we consider parameter estimation methods that are characterized by minimizing a cost function $V(\theta)$:

$$V(\theta) := \frac{1}{2}\boldsymbol{\varepsilon}(\theta)^T P_v^{-1}\boldsymbol{\varepsilon}(\theta), \tag{3}$$

where the prediction error sequence $\boldsymbol{\varepsilon}$ is defined as

$$\boldsymbol{\varepsilon}(\theta) = \mathbf{y} - \hat{\mathbf{y}} = \mathbf{y} - h(\theta, \mathbf{u}; x_0), \tag{4}$$

where $\mathbf{y}$ denotes the measured output sequence and $\hat{\mathbf{y}}$ the predictor sequence, and P_v is a weighting matrix that could represent (an estimate of) the covariance matrix of the noise sequence $\mathbf{v}$ that is supposed to act on the measured output. In the rest of the chapter the shorthand notation $\hat{\mathbf{y}}(\theta)$ is used to indicate $h(\mathbf{u}, \theta; x_0)$.

The Jacobian of $V(\theta)$ with respect to the parameters is

$$\frac{\partial V(\theta)}{\partial \theta} = \frac{\partial \boldsymbol{\varepsilon}(\theta)^T}{\partial \theta} P_v^{-1}\boldsymbol{\varepsilon}(\theta) = -\frac{\partial \hat{\mathbf{y}}(\theta)^T}{\partial \theta} P_v^{-1}(\mathbf{y} - \hat{\mathbf{y}}(\theta)). \tag{5}$$

The Hessian of $V(\theta)$ with respect to the parameters is

$$\frac{\partial^2 V(\theta)}{\partial \theta^2} = \frac{\partial \boldsymbol{\varepsilon}(\theta)^T}{\partial \theta} P_v^{-1}\left(\frac{\partial \boldsymbol{\varepsilon}(\theta)^T}{\partial \theta}\right)^T + S = \frac{\partial \hat{\mathbf{y}}(\theta)^T}{\partial \theta} P_v^{-1}\left(\frac{\partial \hat{\mathbf{y}}(\theta)^T}{\partial \theta}\right)^T + S, \tag{6}$$

where S denotes the second-order information in $\frac{\partial^2 V(\theta)}{\partial \theta^2}$. The Jacobian and Hessian are for a given θ and a given operating point (given by $\mathbf{u}$ and x_0). Parameter estimation now consists in finding the parameter estimate as a minimizing argument of the cost function $V(\theta)$

$$\hat{\theta} := \arg\min_{\theta} V(\theta). \tag{7}$$

At $\hat{\theta}$ the cost function $V(\theta)$ is minimized and the Jacobian (5) at $\hat{\theta}$ is zero, i.e. $\frac{\partial V(\theta)}{\partial \theta} = 0$ at $\hat{\theta}$.

3.2 Analyzing local identifiability in $\hat{\theta}$

Local identifiability in $\hat{\theta}$ is generally evaluated by the test whether the optimization problem (7) has a unique solution in the parameter space. By locally approximating the cost function $V(\theta)$ by a quadratic function[2] (and thus neglecting the second

---

[2] This is achieved by approximating $\hat{\mathbf{y}}(\theta)$ with a first-order Taylor expansion around $\hat{\theta}$.

order term S in (6)), uniqueness of $\hat{\theta}$ is guaranteed if the Hessian at $\hat{\theta}$ is positive definite, i.e. $\frac{\partial^2 V(\theta)}{\partial \theta^2} > 0$ at $\hat{\theta}$, which in this case is equivalent to rank $\frac{\partial^2 V}{\partial \theta^2} = q$. This is a sufficient condition for local identifiability in $\hat{\theta}$, see e.g. [3], [9] and [15].

The considered rank test is naturally performed by applying a singular value decomposition (SVD):

$$\frac{\partial^2 V(\theta)}{\partial \theta^2} = U \Sigma V^T = \begin{bmatrix} U_1 & U_2 \end{bmatrix} \begin{bmatrix} \Sigma_1 & 0 \\ 0 & 0 \end{bmatrix} \begin{bmatrix} V_1^T \\ V_2^T \end{bmatrix},$$

where matrices U and V are unitary matrices, $\Sigma_1 = \text{diag}(\sigma_1, \ldots, \sigma_p)$ with $\sigma_1 \geq \cdots \geq \sigma_p$.

If $p = q$ then identifiability is confirmed. If $p < q$ then the column space of U_1 represents the subspace of the parameter space that is identifiable, and the column space of U_2 is its orthogonal complement, characterizing the subspace that is not identifiable.

As a result, the SVD of the Hessian can be used to extend the qualitative treatment of the question whether or not a particular model structure is identifiable, to a quantitative property of specifying the identifiable parameter space. The columns of U_1 basically act as basis functions in the parameter space, determining the linear combinations of the original parameters that will be identifiable from the measurements. Differently formulated, this would point to a reparametrization of the model structure by defining a reduced order parameter $\rho \in \mathbb{R}^p$ defined by

$$\theta = U_1 \rho \tag{8}$$

leading to an identifiable model structure in the parameter ρ. The attractive feature of this mapping is that it allows to identify ρ while the estimated result $\hat{\rho}$ can be uniquely interpreted in terms of the original physical parameters θ through the mapping (8). The limitation of the approach is of course that only *linear* parameter transformations are considered.

3.3 Approximating the identifiable parameter space

When in the SVD of the Hessian singular values are found that are (very) small, this points to directions in the parameter space that have very limited (but nonzero) influence on the cost function V. In identification terms this correspond to directions in the parameter space in which the variance is (very) large. The Hessian evaluated at $\hat{\theta}$ is connected to the variance of $\hat{\theta}$, since for the Gaussian case (and provided that $\hat{\theta}$ is a consistent estimate) it follows that

$$cov\left(\hat{\theta}\right) = J^{-1}$$

with J the Fisher information matrix

$$J = \mathbb{E}\left[\left.\frac{\partial^2 V(\theta)}{\partial \theta^2}\right|_{\hat{\theta}}\right],$$

(9)

where $\mathbb{E}$ denotes expectation ([15]).

We are interested in specifying that part of the parameter space that is best identifiable by removing the subspace that has only a very small influence on the cost
function V. This reasoning would point to removing those parameter (combinations)
from the model structure for which the variance is very large, as was also addressed
in [26] for nonlinear parameter mappings, and in [17] for single parameters.

The essential information on the SVD of the Hessian is now obtained from:

$$\frac{\partial \hat{y}(\theta)^T}{\partial \theta} P_v^{-\frac{1}{2}} = [\, U_1 \;\; U_2 \,] \begin{bmatrix} \Sigma_1 & 0 \\ 0 & \Sigma_2 \end{bmatrix} \begin{bmatrix} V_1^T \\ V_2^T \end{bmatrix}$$

(10)

where the separation between Σ_1 and Σ_2 is chosen in such a way that the singular
values in Σ_2 are considerably smaller than those in Σ_1.

If we now reparametrize the model structure by employing the reduced parameter
ρ determined by $\theta = U_1 \rho$, we have realized a model structure approximation, in
which the parameters to be identified are well identifiable with a limited variance
and the physical interpretation of the parameters remains untouched. The singular
vectors that occur as the columns in U_1 actually can be seen as basis functions in the
parameter space.

With the SVD (10) it follows that the sample estimate of the covariance matrix
of $\hat{\theta}$ becomes:

$$cov(\hat{\theta}) = \begin{cases} [\, U_1 \;\; U_2 \,] \begin{bmatrix} \Sigma_1^{-2} & 0 \\ 0 & \Sigma_2^{-2} \end{bmatrix} \begin{bmatrix} U_1^T \\ U_2^T \end{bmatrix} & \text{for trace}(\Sigma_2) > 0 \\ U_1 \Sigma_1^{-2} U_1^T & \text{for } \Sigma_2 = 0 \end{cases}$$

(11)

while the sample estimate of the covariance matrix of the reparametrized parameter
estimate $U_1 \hat{\rho}$ is given by

$$cov(U_1 \hat{\rho}) = U_1 \Sigma_1^{-2} U_1^T.$$

(12)

This shows that if $\Sigma_2 = 0$ there is no benefit of the reparametrization in terms of
variance of the estimated parameter $\hat{\theta}$. However if nonzero singular values are dis-
carded in Σ_2, i.e. if trace$(\Sigma_2) > 0$ then

$$cov(\hat{\theta}) > cov(U_1 \hat{\rho})$$

showing a covariance that is reduced by the reparametrization. This reduction is
particularly interesting if Σ_2 contains a (very) large number of small singular values.

4 Parameter scaling in identifiability

The notions of identifiability are defined in such a way that the result is independent of any particular scaling of parameters. A scaling happens when choosing a
particular physical unit for a particular parameter, as e.g. using either [nm] or [m] as
measure of distance. One would also expect that issues around identifiability should
not be dependent on these scalings. Also the analysis and test in section (3.2) is
independent of parameter scaling; a scaling of parameters leads to scaled singular
values, but nonzero singular values remain nonzero after scaling and vice versa.
However when considering the singular value decomposition in the approximating
situation of section (3.3) parameter scaling does have an influence on the numerical
values that occur in Σ_1, Σ_2, and therefore can essentially influence the choice for
separating Σ_1 and Σ_2. This particularly plays a role when the physical parameters
contain different physical quantities. E.g. in the case of [13] the physical parameters
relate to saturations (oil/water percentages) and permeabilities in each separate grid-
block that is a result of spatial discretization. The underlying question of parameter
scaling is then: how to balance the variability in the different physical parameters.

It appears that in the approach presented above the absolute variance of parameters is used as a measure for selection, and as a result the selected parameter space
will become dependent on the chosen parameters scales/units. If it is preferred to
arrive at a selection mechanism that is scaling independent, the relative variance of
parameters is an attractive choice, i.e.

$$cov\left(\Gamma_{\hat{\theta}}^{-1}\hat{\theta}\right)$$

where $\Gamma_{\hat{\theta}} = \text{diag}\left(\,|\hat{\theta}_1|\;\;\ldots\;\;|\hat{\theta}_q|\,\right)$. This motivates the analysis of a scaled Hessian

$$\Gamma_{\hat{\theta}}\,\frac{\partial^2 V(\theta)}{\partial\theta^2}\bigg|_{\hat{\theta}}\,\Gamma_{\hat{\theta}}, \tag{13}$$

related to the scaled Fisher information matrix $\tilde{J}$:

$$\tilde{J} = \mathbb{E}\left[\Gamma_{\hat{\theta}}\,\frac{\partial^2 V(\theta)}{\partial\theta^2}\bigg|_{\hat{\theta}}\,\Gamma_{\hat{\theta}}\right]. \tag{14}$$

The appropriate selection of the identifiable parameter space is then obtained by
applying an SVD according to:

$$\Gamma_{\hat{\theta}}\,\frac{\partial\hat{y}(\theta)^T}{\partial\theta}P_v^{-\frac{1}{2}} = \begin{bmatrix} U_1 & U_2 \end{bmatrix}\begin{bmatrix} \Sigma_1 & 0 \\ 0 & \Sigma_2 \end{bmatrix}\begin{bmatrix} V_1^T \\ V_2^T \end{bmatrix}. \tag{15}$$

Consequences of this parameter scaling are illustrated for some simple examples in
Section 9.

Note that the evaluation of the relative variance of parameter estimates for model
structure selection is also done in classical methods when considering the standard
deviation of an estimated parameter related to the parameter value itself, see e.g.

[15] and [12]. However usually the analysis is performed for parameters separately (e.g. is zero included in the parameter confidence interval?). In the analysis presented here linear combinations of parameters are evaluated, thus focussing on the ratio between the lengths of the principle axes of the uncertainty ellipsoids representing the parameter confidence bounds for $\hat{\theta}$. The parameter uncertainty ellipsoid is expressed by

$$\mathscr{D}(\alpha,\hat{\theta}) = \{\theta \mid N(\theta - \hat{\theta})^T P^{-1}(\theta - \hat{\theta}) \leq \chi_{q,\alpha}^2\}$$

with $P = J^{-1}$ the covariance matrix of the estimator, and $\chi_{q,\alpha}^2$ corresponds to a probability level α in the χ_q^2-distribution with q degrees of freedom. This parameter uncertainty ellipsoid is used to specify that $\theta_0 \in \mathscr{D}(\alpha,\hat{\theta})$ with probability α.

5 Relation with controllability and observability

In this section we will show how the identifiable parameter space that results from (10) is related to properties of controllability and observability.

We consider a strictly proper deterministic linear time-varying (LTV) model in discrete-time state-space form, that could result from linearizing a nonlinear model in the vicinity of a nominal trajectory. The model is given by

$$\begin{aligned} x_{k+1} &= A_k(\theta)x_k + B_k(\theta)u_k \\ \hat{y}_k(\theta) &= C_k(\theta)x_k, \end{aligned}$$

where subscript k denotes the time index. The sensitivity of the predicted outputs with respect to the parameter vector θ is element-wise given by

$$\frac{\partial \hat{y}_k(\theta)}{\partial \theta(i)} = C_k(\theta)\frac{\partial x_k}{\partial \theta(i)} + \frac{\partial C_k(\theta)}{\partial \theta(i)}x_k,$$

where $\frac{\partial x_k}{\partial \theta(i)}$ is determined by

$$\frac{\partial x_{k+1}}{\partial \theta(i)} = A_k(\theta)\frac{\partial x_k}{\partial \theta(i)} + \underbrace{\frac{\partial A_k(\theta)}{\partial \theta(i)}x_k + \frac{\partial B_k(\theta)}{\partial \theta(i)}u_k}_{:=\tilde{u}_k^{\theta(i)}}. \tag{16}$$

Without loss of generality we assume that $\frac{\partial C_k}{\partial \theta(i)} = 0$, since C_{k+1} can be made independent of θ by redefining the state. Note that the effect of a parameter change is weighted by the value of current state and input, i.e. in (16) $\frac{\partial A_k(\theta)}{\partial \theta(i)}$ is weighted by x_k and $\frac{\partial B_k}{\partial \theta(i)}$ is weighted by u_k. This means that given a specific model structure, outputs are more sensitive to parameters associated with states that have a large value. In stacked form we can write

$$
\left(\frac{\partial \hat{y}(\theta)^T}{\partial \theta} \right)^T =
$$

$$
= \underbrace{\begin{bmatrix} C_1 & & & 0 \\ & C_2 & & \\ & & \ddots & \\ 0 & & & C_N \end{bmatrix} \begin{bmatrix} I & 0 & & \cdots & \\ A_1 & I & & & \\ A_2 A_1 & A_2 & I & & \\ \vdots & & & \ddots & \vdots \\ A_{N-2}\ldots A_1 & A_{N-2}\ldots A_2 & \cdots & A_{N-2} & I \\ A_{N-1}\ldots A_1 & A_{N-1}\ldots A_2 & \cdots & \cdots & A_{N-1} & I \end{bmatrix}}_{\tilde{\mathscr{O}} \in \mathbb{R}^{N(p\times n)}} \times
$$

$$
\underbrace{\begin{bmatrix} \tilde{u}_0^{\theta(1)} & \cdots & \tilde{u}_0^{\theta(i)} & \cdots & \tilde{u}_0^{\theta(q)} \\ \tilde{u}_1^{\theta(1)} & \cdots & \tilde{u}_1^{\theta(i)} & \cdots & \tilde{u}_1^{\theta(q)} \\ \vdots & & \vdots & & \vdots \\ \tilde{u}_{N-1}^{\theta(1)} & \cdots & \tilde{u}_{N-1}^{\theta(i)} & \cdots & \tilde{u}_{N-1}^{\theta(q)} \end{bmatrix}}_{\tilde{U} \in \mathbb{R}^{Nn\times q}}, \quad (17)
$$

where we have defined $\tilde{\mathscr{O}}$ and $\tilde{U}$. For a change in the model parameters the term $\tilde{u}_k^{\theta(i)}$ in (16) was given by $\frac{\partial A(\theta)}{\partial \theta(i)} x_k + \frac{\partial B(\theta)}{\partial \theta(i)} u_k$. With this expression, equation (17) provides an appealing interpretation of the Jacobian of the predicted outputs with respect to the parameter vector: a change in any of the parameters is translated into a perturbation of the state of the system, which propagates over time through the dynamical system to reveal the effect on the predicted outputs. As a result the Jacobian is seen to be determined by three factors: the current state and input (x_k, u_k), secondly, the mapping from a model parameter perturbation to a state change (sensitivities of A and B), and thirdly the mapping from a state perturbation to a change in the output (observability properties in $\tilde{\mathscr{O}}$). Indeed only parameter changes that result in state perturbations contained in the column space of $(\tilde{\mathscr{O}})$ can be identified. Moreover, the current state and input need to be nonzero. In the situation that the initial state $x_0 = 0$, and $\frac{\partial B}{\partial \theta} u = 0$, it follows that the state will be significant only in the controllable state space. Note, however, that the initial state x_0 and natural disturbances of the states can also contribute to x being nonzero, which would allow model parameters to be identified.

6 Cost function minimization in identification

In this section we will show how identifiability properties of the model structure appear in gradient-based iterative parameter estimation algorithms. If we iteratively solve for a parameter estimate $\hat{\theta}$ by minimizing a cost function $V(\theta)$, the general update rule in step m of a Newton-type algorithm is given by

$$\hat{\theta}_{m+1} = \hat{\theta}_m - \gamma \left(\frac{\partial^2 V}{\partial \theta^2} \right)^{-1} \frac{\partial V}{\partial \theta}, \tag{18}$$

where γ denotes a scalar damping factor. Note that in this expression the partial derivatives are evaluated in the local parameter $\hat{\theta}_m$. In contrast with the analysis in the previous section this local parameter does not necessarily reflect a (local) minimum of the cost function V.

If we consider the prediction error cost function as used before, then for the model structure considered and after linearization of $\hat{y}(\theta)$ around parameter θ_m the update rule becomes

$$\hat{\theta}_{m+1} = \hat{\theta}_m + \gamma \left(\frac{\partial \hat{y}(\theta)^T}{\partial \theta} \left(\frac{\partial \hat{y}(\theta)^T}{\partial \theta} \right)^T \right)^{-1} \frac{\partial \hat{y}(\theta)^T}{\partial \theta} (\mathbf{y} - \hat{\mathbf{y}}(\theta)). \tag{19}$$

where P_v is considered identity for notational simplicity.
The parameter update (19) is actually a Gauss-Newton step ([5]), employing a first order Taylor expansion of $\hat{y}(\theta)$ around θ_m, similar to the approximation in section 3.3. As an alternative, a Steepest-Descent algorithm ([5]) approximates the Hessian with any positive definite matrix, where standard the identity matrix is chosen. As a result, the update rule in the considered situation becomes:

$$\hat{\theta}_{m+1} = \hat{\theta}_m + \gamma \frac{\partial \hat{y}(\theta)^T}{\partial \theta} (\mathbf{y} - \hat{\mathbf{y}}(\theta)).$$

If the model structure is not identifiable in $\hat{\theta}_m$ the matrix inverse in (19) will not exist. Although this is often indicated as a serious problem for iterative optimization algorithms it can simply be overcome by restricting the update rule to make steps only in that part of the parameter space that does influence the output predictor, see e.g. [18, 28, 19]. This actually come down to utilizing the pseudo-inverse of the Jacobian in (19), on the basis of the SVD:

$$\frac{\partial \hat{y}(\theta)^T}{\partial \theta} = \begin{bmatrix} U_1 & U_2 \end{bmatrix} \begin{bmatrix} \Sigma_1 & 0 \\ 0 & \Sigma_2 \end{bmatrix} \begin{bmatrix} V_1^T \\ V_2^T \end{bmatrix} \tag{20}$$

with $\Sigma_1 \in \mathbb{R}^{p \times p}$. If $\Sigma_2 = 0$, the update rule for the Gauss-Newton iteration can then be replaced by

$$\hat{\theta}_{m+1} = \hat{\theta}_m + \gamma U_1 \Sigma_1^{-1} V_1^T (\mathbf{y} - \hat{\mathbf{y}}(\theta)),$$

while the update rule of Steepest-Descent is given by

$$\hat{\theta}_{m+1} = \hat{\theta}_m + \gamma U_1 \Sigma_1 V_1^T (\mathbf{y} - \hat{\mathbf{y}}(\theta)).$$

Both algorithms update the parameter only in the subspace that is determined by the column space of U_1, being the locally identifiable subspace of the parameter space in the local point $\hat{\theta}_m$.

Note that the difference between the two update mechanisms is that Steepest-Descent emphasizes the vectors of U_1 that correspond to large singular values of the Jacobian, while Gauss-Newton emphasizes the vectors of U_1 that correspond to small singular values of the Jacobian.

Large singular values of the Jacobian are associated with directions in which the predictions are very sensitive to a change in the parameters. Indeed, Steepest-Descent looks for the direction in which the cost function decreases with the least amount of effort in changing the parameters. Gauss-Newton follows an opposite strategy. Here the algorithm looks for a change in predicted outputs (i.e. cost function) which is induced by the largest change in the parameters.

It can simply be verified that parameter scaling does influence the estimate of the steepest-descent algorithm, in contrast with the Gauss-Newton algorithm which is scaling invariant, see e.g. [5]. This scaling-invariance however is only true if the Hessian has full rank in $\theta = \theta_m$.

Similar to the analysis in the previous sections the rank reduction of the Jacobian, as represented in (20) can of course be enforced if the SVD shows a large number of small singular values in Σ_2, and the Jacobian is approximated by setting $\Sigma_2 = 0$.

A similar approach of Jacobian reduction is employed in the fully parametrized state-space model identification using so-called data-driven local coordinates of [18, 19] as well as in subspace identification [28], where search directions are chosen to be orthogonal to the tangent space of the manifold representing equivalent models. See also [31] for a further comparison of methods. If the main interest of the modelling procedure is to identify (linear) system dynamics, these approaches are attractive as they simply use the parameters as vehicles to arrive at an appropriate system model. However, in this paper we aim at preserving the physical interpretation of the parameters and therefore are more focussing on the uniqueness of the parameters estimates in order to obtain reliable long-term (non-linear) model predictions.

7 A Bayesian approach

Lack of identifiability of a model structure and the subsequent non-uniqueness of parameters that are estimated on the basis of measurement data, can be dealt with in different ways. One way is to reduce the parameter space in the model structure, as indicated in the previous sections. Alternatively additional prior information can be added to the identification problem. In those situations where a parameter estimate may not be uniquely identifiable from the data, a regularization term can be added to the cost function that takes account of prior knowledge of the parameters to be estimated. In this setting an alternative (Bayesian) cost function is considered:

$$V_p(\theta) := V(\theta) + \frac{1}{2}(\theta - \theta_p)P_{\theta_p}^{-1}(\theta - \theta_p), \tag{21}$$

where the second term represents the weighted mismatch between the parameter vector and the prior parameter vector θ_p with covariance P_{θ_p}. When again the model output $\hat{\mathbf{y}}(\theta)$ is approximated using a first-order Taylor expansion around θ_p, the Hessian of (21) becomes:

$$\frac{\partial^2 V_p(\theta)}{\partial \theta^2} = \frac{\partial \hat{\mathbf{y}}(\theta)^T}{\partial \theta} P_v^{-1} \left(\frac{\partial \hat{\mathbf{y}}(\theta)^T}{\partial \theta} \right)^T + P_{\theta_p}^{-1}. \tag{22}$$

Since $P_{\theta_p}^{-1}$ is positive definite by construction and the first term is positive semi-definite, the Hessian has full rank and the parameter estimate

$$\hat{\theta}_{bayes} = \arg\min_\theta V_p(\theta)$$

is unique. This uniqueness is guaranteed by the prior information that has been added to the problem. Formally there can still be lack of identifiability, however it is not any more reflected in a non-unique parameter estimate. A consequence of this approach is that the obtained parameter estimate may be highly influenced by the prior information, and less by the measurement data.

The covariance matrix of the Bayesian parameter estimate can also be analyzed using the classical prediction error theory, see [15]. Under ideal circumstances (consistent estimation and $\theta_p = \theta_0$ (!)) it can be shown that

$$cov(\hat{\theta}_{bayes}) = \left[\mathbb{E} \left. \frac{\partial^2 V_p(\theta)}{\partial \theta^2} \right|_{\theta_0} \right]^{-1}. \tag{23}$$

In other words, the inverse of the Hessian of the identification criterion remains to play the role of (sample estimate of) the parameter covariance matrix, and the same considerations as discussed in the earlier sections can be applied to the SVD analysis of this Hessian. By appropriately operating on the expression for the Hessian (22), it can be shown that a relevant SVD analysis for dimension reduction can now be applied to

$$P_{\theta_p}^{\frac{T}{2}} \frac{\partial \hat{\mathbf{y}}(\theta)^T}{\partial \theta} P_v^{-\frac{1}{2}},$$

which in [23] is referred to as the dimensionless sensitivity matrix.

It may be clear that the parameter estimate becomes highly dependent on the prior information, and that bias will occur when the parameter prior θ_p is not correct.

It has to be noted that this Bayesian approach is typically followed when using sequential estimation algorithms for joint parameter and state estimation, as in Extended Kalman Filters and variations thereof, such as the Ensemble Kalman Filter, see e.g. [8].

8 Structural identifiability

The question whether parameters can be uniquely identified from data basically consists of two parts. The first part concerns the model structure: is it possible at all to distinguish two given parameters, provided that the input is chosen in the best possible way? This property is called structural identifiability of a model structure. The second part concerns the issue whether the actual input is informative enough to allow this distinction. In the previous sections both parts were considered simultaneously. In this section only the first part is investigated. Consider definition 2 on structural identifiability. Without loss of generality, but for ease of notations, we will limit attention to the SISO case. The multivariable case is also treated in [27]. Note that $G(z, \theta)$ can be written as:

$$G(z, \theta) = \sum_{k=1}^{\infty} M(k, \theta) z^{-k}, \tag{24}$$

where $M(k, \theta)$ are the Markov parameters. Based on (24) we argue that equality of $G(z, \theta_1)$ and $G(z, \theta_2)$ is related to equality of the Markov parameters of $G(z, \theta_1)$ and $G(z, \theta_2)$. We can now use the following proposition ([9], [11], [20], [27]):

Proposition 3 *Consider the map* $S_N(\theta) : \Theta \subset \mathbb{R}^q \to \mathbb{R}^N$ *defined by:*

$$S_N(\theta) := [M(1, \theta) \ \dots \ M(N, \theta)]^T. \tag{25}$$

Then the model structure is locally structural identifiable in θ_m *if* rank $\left(\frac{\partial S_N^T(\theta)}{\partial \theta} \right) = q$ *in* $\theta = \theta_m$.

Both the qualitative question of structural identifiability, and the determination of the "best" structurally identifiable subspace of parameters can now be examined by applying an SVD to the matrix

$$\frac{\partial S_N^T(\theta)}{\partial \theta} \tag{26}$$

and examining the column space of this matrix, see [27]. However also in this problem we need to take care that our (approximate) identifiability test is not dependent on user-chosen parameter scaling, and so we need a premultiplication of (26) with the scaling matrix Γ_{θ_m}. If a parameter has high impact on a particular Markov parameter, but the Markov parameter itself has a very small value, the considered parameter is still a good candidate to be removed in our model structure approximation problem. Therefore an additional weighting of (26) is desired that takes account of the values of the Markov parameters. As a result we consider the column space of the matrix

$$\Gamma_{\theta_m} \frac{\partial S_N^T(\theta)}{\partial \theta} \Gamma_S \tag{27}$$

where for the SISO case $\Gamma_S := \mathrm{diag} \left(|M_1| \ \dots \ |M_N| \right)$. The consequence is that Markov parameters that have a high value are considered to be more important to

include than Markov parameters with a small value.[3]
The row space of (27) that relates to the dominant singular values of the matrix, now is a representation of the parameter space of the approximated model structure. The Jacobian matrix (26) can be calculated analytically, as is shown in [27] for state space model structures.

The structurally identifiable problem and the identifiability problem are of course closely related to each other. This can be observed by realizing that

$$\frac{\partial \hat{y}(\theta)^T}{\partial \theta} = \frac{\partial S_N^T(\theta)}{\partial \theta} \Phi_N, \tag{28}$$

where Φ_N is given by

$$\Phi_N = \begin{bmatrix} u_1 & u_2 & \cdots & u_N \\ & u_1 & & \vdots \\ & & \ddots & \\ & & & u_1 \end{bmatrix} \tag{29}$$

and the derivatives are evaluated at $\theta = \theta_m$.
Note that the matrix Φ_N with input signals acts as a weighting matrix in (28) in a similar way as the weighting matrix Γ_S does in (27).

9 Examples

In order to illustrate the concepts, and in particular the role of the scaling/weighting functions, we will now discuss two examples where we have chosen a very simple SISO finite impulse response (FIR) model. The model structure will be approximated using the previously discussed identifiability analysis, where we assume that $P_v = I$.

Example 4 *Consider the data-generating system*

$$y(t) = \alpha_0 u(t-1) + \beta_0 u(t-2)$$

with $\alpha_0 = 10^6$ and $\beta_0 = 10^{-6}$, and $\theta_0 := [\alpha_0 \ \beta_0]^T$. Consider the input/output model structure

$$y(t,\theta) = \alpha u(t-1) + \beta u(t-2), \quad \theta := [\alpha \ \beta]^T.$$

For an analysis of the local identifiability in $\theta = \theta_0$ we consider

$$\psi(t,\theta_0) := \frac{\partial y(t,\theta)}{\partial \theta} = \begin{bmatrix} u(t-1) \\ u(t-2) \end{bmatrix} \tag{30}$$

so that the Fisher information matrix J is given by

---

[3] Note that in a more generalized setting this weighting should be replaced by a weighting that takes account of the application in which the model is used.

$$J = \mathbb{E} \sum_{t=1}^{N} \psi(t,\theta_0)\psi(t,\theta_0)^T$$

$$= N\mathbb{E}\begin{pmatrix} u(t-1) \\ u(t-2) \end{pmatrix}\begin{pmatrix} u(t-1) \\ u(t-2) \end{pmatrix}^T = N\begin{bmatrix} R_u(0) & R_u(1) \\ R_u(1) & R_u(0) \end{bmatrix}$$

with $R_u(\tau) = \mathbb{E}[u(t)u(t-\tau)]$.

The scaled Fisher information matrix $\tilde{J}$ of (14) for a local analysis around θ_0 is then given by

$$N\begin{bmatrix} \alpha_0 & 0 \\ 0 & \beta_0 \end{bmatrix}\begin{bmatrix} R_u(0) & R_u(1) \\ R_u(1) & R_u(0) \end{bmatrix}\begin{bmatrix} \alpha_0 & 0 \\ 0 & \beta_0 \end{bmatrix}.$$

The relative parameter variance is indicated by $\tilde{J}^{-1}$. In the case of a unit variance white noise input, it follows that

$$\tilde{J} = N\begin{bmatrix} 10^{12} & 0 \\ 0 & 10^{-12} \end{bmatrix},$$

while the unscaled Fisher information matrix satisfies $J = N \cdot I$. Analysis of $\tilde{J}$ shows that the second parameter can very well be neglected, leading to an approximate model structure $y(t) = \alpha u(t-1)$.

Structural identifiability analysis without scaling shows that both parameters are structurally identifiable, since $\frac{\partial S_N^T(\theta)}{\partial \theta} = I$. However, including both scaling matrices Γ_θ and Γ_S, we obtain

$$\Gamma_\theta \frac{\partial S_N^T(\theta)}{\partial \theta}\Gamma_S = \begin{bmatrix} 10^{12} & 0 \\ 0 & 10^{-12} \end{bmatrix},$$

also showing that the second parameter can be very well neglected. In light of section 5 we remark that in $\theta = \theta_0$ this model is poorly observable/controllable and as a result it is also poorly identifiable.

Example 5 *In this example the same data-generating system as in the previous example is considered. Consider the input/output model structure*

$$y(t,\theta) = \alpha u(t-1) + 10^{-6}\gamma u(t-2), \quad \theta := [\alpha\ \gamma]^T.$$

where $\gamma_0 = 1$. In comparison with the previous example we have scaled the second parameter with a factor 10^{-6}. This can be thought of to be the result of choosing a different physical unit for the parameter. The scaled Fisher information matrix $\tilde{J}$ of (14) is

$$N\begin{bmatrix} \alpha_0 & 0 \\ 0 & \gamma_0 \end{bmatrix}\begin{bmatrix} R_u(0) & 10^{-6}R_u(1) \\ 10^{-6}R_u(1) & 10^{-12}R_u(0) \end{bmatrix}\begin{bmatrix} \alpha_0 & 0 \\ 0 & \gamma_0 \end{bmatrix}.$$

Under the same input conditions it follows that

$$\tilde{J} = N\begin{bmatrix} 10^{12} & 0 \\ 0 & 10^{-12} \end{bmatrix},$$

while the unscaled Fisher information matrix is

$$J = N \begin{bmatrix} 1 & 0 \\ 0 & 10^{-12} \end{bmatrix}.$$

Whereas the unscaled matrix is essentially different from the previous example, the scaled analysis shows again that the second parameter can very well be neglected and that the model structure can be approximated with $y(t) = \alpha u(t-1)$.
Structural identifiability analysis without scaling shows that α is structurally best identifiable, since

$$\frac{\partial S_N^T(\theta)}{\partial \theta} = \begin{bmatrix} 1 & 0 \\ 0 & 10^{-6} \end{bmatrix}.$$

Including both scaling matrices Γ_θ and Γ_S, we obtain in quadratic form

$$\Gamma_\theta \frac{\partial S_N^T(\theta)}{\partial \theta} \Gamma_S = \begin{bmatrix} 10^{12} & 0 \\ 0 & 10^{-12} \end{bmatrix},$$

being exactly the same as matrix as in the previous example, meaning that the structural identifiability analysis is now-scaling invariant.

The examples are of course very simple, and they are meant to illustrate the basic phenomena that might occur in large scale physical structures. Use of a notion of relative variance, reflected in a scaled Fisher information matrix, leads to selection results that are scaling-invariant. The fact that we consider a local analysis only is of course a limitation of the results presented here.

10 Conclusions

The question whether a large scale (nonlinear) physical model structure is identifiable, is usually considered in a qualitative way. In this chapter the notion of identifiability is quantified and it is shown how the model structure can be approximated so as to achieve identifiability, while retaining the interpretation of the physical parameters. In this chapter this question has been addressed in a prediction error setting. The analysis has been related to iterative optimization algorithms (like Gauss-Newton and Steepest-Descent) and to Bayesian estimation. It has been shown that parameter scaling becomes relevant when approximating model structures.

Acknowledgements This chapter is dedicated to Okko Bosgra, a great colleague and supervisor, at the occasion of his 65th birthday, and with great appreciation for the many stimulating discussions we have had over the years.

References

[1] Bai, E.W.: An optimal two-stage identification algorithm for hammerstein-wiener nonlinear systems. Automatica **34**(3), 333–338 (1998)

[2] Bamieh, B., Giarré, L.: Identification of linear parameter varying models. Int. J. Robust and Nonlinear Control **12**, 841–853 (2002)

[3] Bellman, R., Åström, K.J.: On structural identifiability. Mathematical Biosciences **7**, 329–339 (1970)

[4] Berntsen, H.E., Balchen, J.G.: Identifiability of linear dynamical systems. In: P. Eykhoff (ed.) Proc. 3rd IFAC Symp. Identification and System Parameter Estimation, pp. 871–874. The Hague (1973)

[5] Dennis, J., Schnabel, R.: Numerical Methods for Unconstrained Optimization and Nonlinear Equations. Prentice-Hall, Englewood Cliffs, New Jersey (1983)

[6] Dötsch, H.G.M., Van den Hof, P.M.J.: Test for local structural identifiability of high-order non-linearly parameterized state space models. Automatica **32**(6), 875–883 (1996)

[7] Evans, N., Chapman, M., Chappell, M., Godfrey, K.: Identifiability of uncorrelated nonlinear rational systems. Automatica **38**, 1799–1805 (2002)

[8] Evensen, G.: Data Assimilation. Springer (2007)

[9] Glover, K., Willems, J.C.: Parameterizations of linear dynamical systems: canonical forms and identifiability. IEEE Trans. Automatic Control **19**(6), 640–646 (1974)

[10] Godfrey, K.R.: Compartmental Models and their Application. Academic Press, London (1983)

[11] Grewal, M.S., Glover, K.: Identifiability of linear and nonlinear dynamical systems. IEEE Trans. Automatic Control **21**(6), 833–837 (1976)

[12] Hjalmarsson, H.: From experiment design to closed-loop control. Automatica **41**(3), 393–438 (2005)

[13] Jansen, J.D., Bosgra, O.H., Van den Hof, P.M.J.: Model-based control of multiphase flow in subsurface oil reservoirs. J. Process Control **18**, 846–855 (2008)

[14] Lee, L.H., Poolla, K.: Identification of linear parameter-varying systems using nonlinear programming. J. of Dynamic Systems, Measurement and Control **121**(1), 71–78 (1999)

[15] Ljung, L.: System Identification: Theory for the User, 2nd edition. Prentice-Hall, Englewood Cliffs, NJ (1999)

[16] Ljung, L., Glad, T.: On global identifiability for arbitrary model parameterizations. Automatica **30**(2), 265–276 (1994)

[17] Lund, B.F., Foss, B.A.: Parameter ranking by orthogonalization - applied to nonlinear mechanistic models. Automatica **44**, 278–281 (2008)

[18] McKelvey, T., Helmersson, A.: System parameterization of an overparameterized model class: improving the optimization algorithm. In: 36th IEEE Conf. Decision and Control, pp. 2984–2989. San Diago, CA, USA (1997)

[19] McKelvey, T., Helmersson, A., Ribarits, T.: Data driven local coordinates for multivariable linear systems and their application to system identification. Automatica **40**, 1629–1635 (2004)

[20] Norton, J.P.: Normal-mode identifiability analysis of linear compartmental systems in linear stages. Mathematical Biosciences **50**, 95–115 (1980)

[21] Pintelon, R., Schoukens, J.: System Identification: A Frequency Domain Approach. IEEE Press, Piscataway, NJ (2001)

[22] Stigter, J.D., Peeters, R.L.M.: On a geometric approach to the structural identifiability problem and its application in a water quality case study. In: Proc. European Control Conf. Koss, Greece (2007)

[23] Tavakoli, R., Reynolds, A.C.: History matching with parameterization based on the SVD of a dimensionless sensitivity matrix. In: Proc. SPE Reservoir Simulation Symposium, Texas, USA (2009)

[24] Tóth, R.: Modeling and identification of linear parameter-varying systems - an orthornomal basis function approach. Ph.D. thesis, Delft Univ. Technology, Delft, The Netherlands (2008)

[25] Tóth, R., Heuberger, P.S.C., Van den Hof, P.M.J.: LPV system identification with globally fixed orthonormal basis functions. In: Proc. 46th IEEE Conf. Decision and Control, pp. 3646–3653. New Orleans, LA (2007)

[26] Vajda, S., Rabitz, H., Walter, E., Lecourtier, Y.: Qualitative and quantitative identifiability analysis of nonlinear chemical kinetic models. Chem. Eng. Comm. **83**, 191–219 (1989)

[27] Van Doren, J.F.M., Van den Hof, P.M.J., Jansen, J.D., Bosgra, O.H.: Determining identifiable parameterizations for large-scale physical models in reservoir engineering. In: M. Chung, P. Misra, H. Shim (eds.) Proc. 17th IFAC World Congress, pp. 11,421–11,426. Seoul (2008)

[28] Verdult, V.: Nonlinear system identification: A state-space approach. Ph.D. thesis, Delft Univ. Technology, Delft, The Netherlands (2002)

[29] Verdult, V., Verhaegen, M.: Subspace identification of multivariable linear parameter-varying systems. Automatica **38**(5), 805–814 (2002)

[30] Walter, E.: Identifiability of Parametric Models. Pergamon Press, Oxford (1987)

[31] Wills, A., Ninness, B.M.: On gradient-based search for multivariable system estimates. IEEE Trans. Automatic Control **53**(1), 298–306 (2008)

[32] van Wingerden, J., Verhaegen, M.: Subspace identification of bilinear and LPV systems for open and closed loop data. Automatica **45** (2009). To appear

[33] Zhu, Y.C.: Estimation of an N-L-N Hammerstein-Wiener model. Automatica **38**(9), 1607–1614 (2002)

[34] Zhu, Y.C., Xu, X.: A method of LPV model identification for control. In: M. Chung, P. Misra, H. Shim (eds.) Proc. 17th IFAC World Congress, pp. 5018–5023. Seoul (2008)

Part III
Applications in Motion Control Systems and Industrial Process Control

Recovering Data from Cracked Optical Discs using Hankel Iterative Learning Control

Maarten Steinbuch, Jeroen van de Wijdeven, Tom Oomen, Koos van Berkel, and George Leenknegt

Abstract Optical discs, including Compact Discs (CDs), Digital Versatile Discs (DVDs), and Blu-ray Discs (BDs), can get cracked during storage and usage. Such cracks commonly lead to discontinuities in the data track, potentially preventing reading of the data on the disc. The aim of the present paper is to improve tracking performance of the optical disc drive in the presence of cracks. A Hankel Iterative Learning Control (ILC) algorithm is presented that can perfectly steer the lens *during* the crack towards the beginning of the track immediately *after* the crack, i.e., the actuator is steered appropriately during the crack crossing to compensate for the discontinuity in the data track. Experimental results confirm improved reading capabilities of cracked discs. The presented approach potentially enables the recovery of data from cracked discs that were previously considered as unreadable.

1 Introduction

Optical discs, including Compact Discs (CDs), Digital Versatile Discs (DVDs), and Blu-ray Discs (BDs), are media with data written on a layer by means of pits and lands in a spiral track, see Figure 1. These optical discs can get cracked in case they have been subjected to a static or dynamic mechanical load over a certain period of time. Examples of such damaged discs arise in vertical disc storage of many

Maarten Steinbuch, Tom Oomen, and Koos van Berkel
Eindhoven University of Technology, Department of Mechanical Engineering, Control Systems Technology group, P.O. Box 513, 5600 MB Eindhoven, The Netherlands, e-mail: m.steinbuch@tue.nl, t.a.e.oomen@tue.nl, k.v.berkel@gmail.com

Jeroen van de Wijdeven
TMC Mechatronics, P.O. Box 700, 5600 AS Eindhoven, The Netherlands, e-mail: jeroen.van.de.wijdeven@tmc.nl

George Leenknegt
Advanced Research Center, Philips Lite-On Digital Solutions Netherlands, Glaslaan 2, 5616 LW Eindhoven, The Netherlands, e-mail: george.leenknegt@pldsnet.com

P.M.J. Van den Hof et al. (eds.), *Model-Based Control: Bridging Rigorous Theory and Advanced Technology*, DOI: 10.1007/978-1-4419-0895-7_9,
© Springer Science + Business Media, LLC 2009

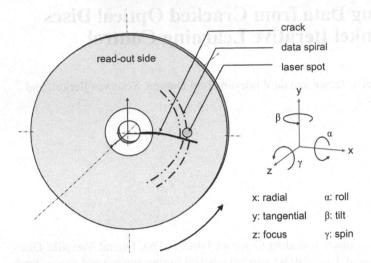

Fig. 1 Schematic top view of a typical cracked optical disc.

discs in a spindle [4] and in high-speed reading of warped or bent discs [14]. The resulting cracks are typically sharp, starting from the center of the disc and are possibly unfinished, see Figures 1, 2, and 3.

The general observation with respect to this specific type of cracked discs is that the amount of lost data, caused by the damaged disc surface or information layer, is only marginal. Specifically, due to data redundancy in the encoded track data, the existing error correction can recover the original data that is contained on the cracked disc. The main problem is a servo control problem: when the discontinuity is particularly large, the laser may lose track in radial direction (track-loss) and/or in focus direction (out-of-focus). The first case results in a continuity failure of the data stream and in the second case the disc gets outside the observation range of the lens [8]. Both consequences are fatal for reading optical discs. At present, control solutions in commercial drives include the so-called defect detector (DEFO) that switches off the normal feedback controller during the passing of a crack and holds the controller output at a constant value.

Although present approaches can cope with cracked discs to a certain extent, the tolerable crack dimensions are limited due to the fact that the zero-order-hold at the input does not result in an optimal connection between the end of the track before the crack and the beginning of the track after the crack. To improve tracking performance immediately after the crack, the lens can be steered towards the beginning of the track during the time the crack is passing. However, the design of an optimal command signal requires future information since it is not known *a priori* where the track is after the crack has passed. Thus, to compensate for the track discontinuity, the properties of the discontinuity should be known in an open-loop type of compensation algorithm.

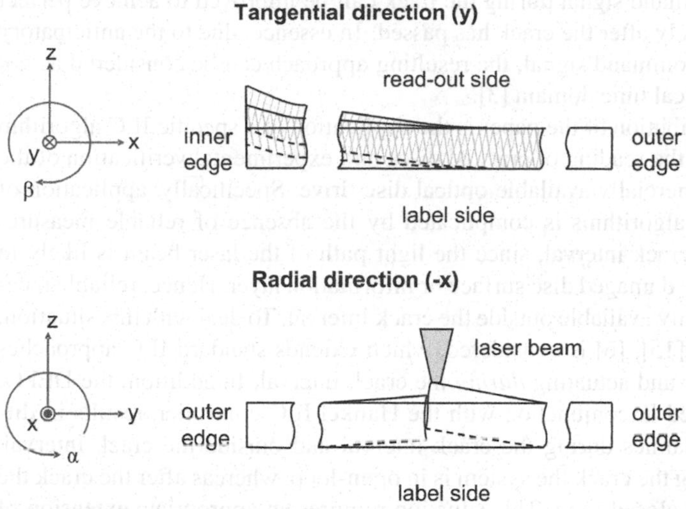

Tangential direction (y)

Fig. 2 Schematic side views of a typical cracked optical disc.

Fig. 3 Photograph of a cracked optical disc.

In the present paper, an Iterative Learning Control (ILC) algorithm [2], [3], [7], [11] is proposed that improves the data recovery in optical cracked discs by implementing a command signal that anticipates on the track location immediately after the crack. The main idea is that the crack results in a, possibly slowly varying, repeating error in case of rotating discs. By using measurement data from previous rotations, the command signal during the track can be improved to achieve perfect tracking immediately after the crack has passed. In essence, due to the anticipatory behaviour of the command signal, the resulting approach can be considered as not causal in the physical time domain [3].

The main contribution of the paper is the application of a specific ILC algorithm that can deal with the reading of cracked discs and experimental verification of the method in a commercially available optical disc drive. Specifically, application of the standard ILC algorithms is complicated by the absence of reliable measurements during the crack interval, since the light path of the laser beam is likely to be disturbed by the damaged disc surface or information layer. Hence, reliable measurement data is only available outside the crack interval. To deal with this situation, Hankel ILC [16], [15], [6] is considered, which extends standard ILC approaches by observing *after* and actuating *during* the crack interval. In addition, the DEFO, that is implemented in conjunction with the Hankel ILC controller, results in different system dynamics during the crack interval and outside the crack interval. Specifically, during the crack the system is in open-loop, whereas after the crack the system is again in closed-loop. This situation requires an appropriate extension of the Hankel ILC framework. The learning controller is implemented in the Digital Signal Processor (DSP) of a Philips BD1-player for experimental evaluation. Experimental results confirm improved performance when reading cracked Compact Disc-Recordables (CD-Rs). Specifically, the Hankel ILC controller is able to compensate discontinuities that are significantly beyond the observation range of the lens while the tracking error does not contain systematic errors after the crack has passed.

This paper is organized as follows. In Section 2 the drive and the cracked disc are introduced. The theory of Hankel ILC is presented in Section 3. In Section 4, implementation aspects are discussed. In Section 5, experimental results are presented. Finally, conclusions are provided in Section 6.

2 Experimental setup

In this section, the relevant system parts are discussed. Specifically, in Section 2.1, the optical storage principle is introduced. Then, in Section 2.2 the class of considered damaged discs is specified. Finally, the motion system is presented in Section 2.3.

2.1 Optical storage principle

The principle of optical storage is schematically depicted in Figure 4. The (cracked) optical disc, depicted in the upper left, is suspended to the turntable motor below.

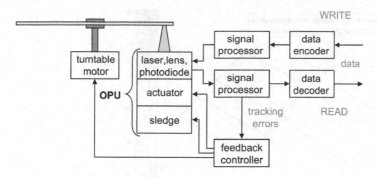

Fig. 4 Schematic representation of the optical storage principle.

The turntable motor rotates the disc with respect to the laser beam in spin direction (γ).

The optical drive extracts data from the optical disc by using a laser-based Optical Pick-up Unit (OPU) to follow the spiral track. The laser beam is generated in the OPU, which typically consists of an actuator, laser, lens, and photodiode. The actuator together with the sledge provide accurate position information of the lens with respect to the disc. The combination of the laser, lens, and photodiode is able to optically read pits and lands in the data spiral on the disc. Additionally, this combination also provides tracking error information, see, e.g., [9], that is required for feedback control and Hankel ILC.

2.2 Cracked disc

In this paper, CD-Rs containing cracks of a certain class are considered with the following properties:

1) the crack is unfinished from the inner to the outer rim in radial direction,
2) the crack is sharp with a limited width in tangential direction (y) of damaged disc surface and information layer,
3) the maximal displacement and rotation of the crack ends is bounded,
4) the relative displacement and rotation of the crack ends between neighboring tracks is bounded,
5) the number of cracks is limited and spatially spread with a certain space between neighboring cracks.

Cracked discs that do not belong to the class characterized by Property 1- 5 can cause problems due to mechanical instability and explosion at high speed, the amount of data loss outside practical error correction capacity, limited actuator range, limited observation range, or confusion between neighboring cracks. Such discs are hence not considered in the present paper. The geometry of a typ-

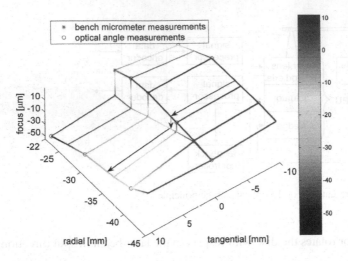

Fig. 5 Measured geometry of a typical cracked disc.

ical crack has been measured using a micrometer and an optical angle meter. The results are depicted in Figure 5, where the arrows indicate the read direction and the encountered discontinuity in the z-direction of a certain track. The measured disc is supposed to satisfy the above Property 1- 5.

2.3 Motion system

The motion system of the considered optical drive, which is a Philips BD1, is a dual-stage combination of an actuator and a sledge, see Figure 4. The optical drive is capable of reading and writing CDs, CD-Rs, DVDs, BDs, etc. In the present research, Hankel ILC is employed to recover data from a cracked CD-R. It is expected that the approach can also be employed to recover data from other optical disc types.

The actuator is designed for dynamic positioning of the lens in focus (z) and radial direction (x), while the sledge provides large radial jumps (x) and offset compensation in roll α, see Figure 1. The actuator is modelled mechanically as a suspended mass with three voice coil motors and has therefore three degrees-of-freedom, i.e., x, z, and β. The tilt β is nominally fixed to zero leaving two degrees-of-freedom. The sledge is modelled as a suspended mass with two electro motors to actuate its two degrees-of-freedom.

To anticipate on the Hankel ILC algorithm in Section 3, it should be noted that ILC requires an approximate model of the plant to update the command signal. Based on the discussion in the previous paragraph, physical models of the actuators are used for both the focus and radial direction, i.e., the model

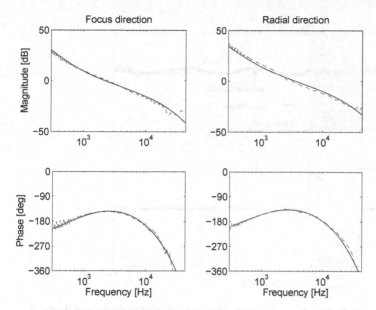

Fig. 6 Bode diagram of the open loop PC frequency response function in focus and radial direction: model P_{sys} (solid), identified frequency response function (dashed).

$$P_{sys} = \frac{k}{s^2 + 2\zeta\omega_n s + \omega_n^2} \tag{1}$$

is considered in both focus and radial direction, where k is a motor constant, ζ is a dimensionless damping coefficient, and ω_n is the undamped natural frequency. In addition, s is a complex indeterminate representing the Laplace variable. Clearly, the model (1) is a continuous time model since the physical system dynamics evolve in the continuous time domain. The model parameters are selected such that the model matches the identified plant frequency response function.

Comparing the model with the identified frequency response function, see Figure 6, where the open-loop gain $P_{sys}C$ is depicted, with C the feedback controller, reveals that the model accurately described the true plant behaviour. In addition, measurements confirm that the plant is approximately decoupled. To facilitate the implementation, independent single-input single-output ILC controllers are designed and implemented. In case of significant interaction in the plant dynamics, both multivariable models and ILC controllers should be considered, see [13] and [16], respectively.

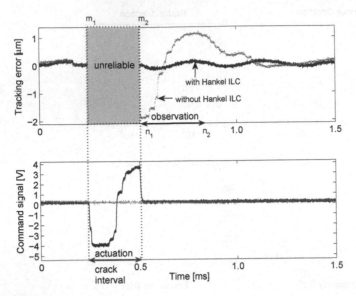

Fig. 7 Timing of actuation and observation: converged situation after application of Hankel ILC, see Section 5 (solid), initial error without Hankel ILC (dotted).

3 Hankel ILC

Iterative learning control is a control technique that iteratively learns an optimal feedforward signal that minimizes tracking errors caused by deterministic disturbances in the same time interval. For the cracked CD, however, the measured tracking error during the duration of the crack is unreliable. Moreover, in the application of the cracked CD, the primary interest is in the design of a feedforward signal *during* the time interval of the crack crossing, such that the focus and radial errors *after* the crack crossing are reduced. In other words, for cracked CDs, the task of the ILC controller is to learn a feedforward signal in one time interval such that the tracking error in the adjacent time interval is reduced. A special form of ILC, referred to as Hankel ILC, is capable of handling this task by exploiting separate time windows for the actuation and observation intervals in the control design, see [16] and Figure 7. The measurement results in Figure 7 already reveal the potential improvement of the tracking error due to Hankel ILC, which will be further explained in Section 5.

3.1 System formulation

Given the discrete-time single-input single-output Linear Time Invariant (LTI) system P with a minimal state-space realization

$$P : \begin{cases} x(t+1) & = & Ax(t) + Bf(t) \\ y(t) & = & Cx(t) + Df(t) \end{cases},$$

(2)

where $f(t)$ is the feedforward signal, $y(t)$ is the measured position, and $t \in \mathbb{Z}$ denotes discrete time. Then $P : f(t) \mapsto y(t)$ for $t = [0, N-1]$ can be written as a convolution matrix equation:

$$\begin{bmatrix} y(0) \\ \vdots \\ y(N-1) \end{bmatrix} = \underbrace{\begin{bmatrix} D & 0 & \cdots & 0 \\ CB & D & \cdots & 0 \\ \vdots & & \ddots & \vdots \\ CA^{N-2}B & \cdots & CB & D \end{bmatrix}}_{P} \begin{bmatrix} f(0) \\ \vdots \\ f(N-1) \end{bmatrix}.$$

(3)

With the actuation interval given by $t \in [m_1, m_2]$, the observation interval by $t \in [n_1, n_2]$, and $n_1 := -m_2 + 1$ (in accordance with Figure 7), the convolutive mapping P_H from $f(t)$ during the actuation interval to $y(t)$ during the observation interval equals

$$\underbrace{\begin{bmatrix} y(n_1) \\ \vdots \\ y(n_2) \end{bmatrix}}_{y} = \underbrace{\begin{bmatrix} CA^{m-1}B & \cdots & CB \\ \vdots & \ddots & \vdots \\ CA^{n+m-2}B & \cdots & CA^{n-1}B \end{bmatrix}}_{P_H} \underbrace{\begin{bmatrix} f(m_1) \\ \vdots \\ f(m_2) \end{bmatrix}}_{f},$$

(4)

$$m = m_2 - m_1 + 1, \quad n = n_2 - n_1 + 1, \quad n_1 = m_2 + 1, \quad (5)$$

with $P_H \in \mathbb{R}^{n \times m}$. In other words, $P_H = W_o P W_a$ with W_o and W_a the observation and actuation time windows, respectively:

$$W_o = \begin{bmatrix} 0_{n \times n_1} & I_n & 0_{n \times (N-n_2-1)} \end{bmatrix},$$

(6)

$$W_a = \begin{bmatrix} 0_{m_1 \times m} \\ I_m \\ 0_{(N-m_2-1) \times m} \end{bmatrix}.$$

(7)

In the common case $\text{rank}(P_H) < \min(m, n)$, the matrix P_H is rank deficient. This rank deficient matrix P_H can be represented as the product of two full rank matrices using a full rank decomposition

$$P_H = P_o P_c,$$

(8)

where $P_o \in \mathbb{R}^{n \times p}$, $P_c \in \mathbb{R}^{p \times m}$, and $p := \text{rank}(P_H)$, representing the following two mappings

$$y = P_o x_{n_1}$$

(9)

$$x_{n_1} = P_c f.$$

(10)

Fig. 8 Hankel ILC scheme.

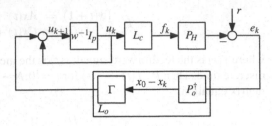

A possible choice regarding P_o and P_c in (9) and (10) is

$$P_o = \begin{bmatrix} C \\ CA \\ \vdots \\ CA^{n-1} \end{bmatrix} \tag{11}$$

$$P_c = \begin{bmatrix} A^{m-1}B & A^{m-2}B & \cdots & B \end{bmatrix}. \tag{12}$$

In (9) and (10), $x_{n_1} \in \mathbb{R}^p$ is the state vector at $t = n_1$. Clearly, in case the state $x_{n_1} = 0$, then in virtue of (9) $y = 0$. Basically, the state x_{n_1} at time $t = n_1$ separates the command signal during the crack and the tracking error after the track, which has a close connection to the Hankel operator, see, e.g., [18]. Hence, the tracking error after the crack is fully determined by the state x_{n_1}.

3.2 Hankel ILC control framework

The Hankel ILC control framework is depicted in Figure 8. The input $f_k \in \mathbb{R}^m$ and output $y_k \in \mathbb{R}^n$ denote the feedforward signal during the actuation time interval and measured output during the observation time interval in trial k, respectively. Trial k in case of reading cracked discs refers to the kth disc revolution after the learning algorithm was switched on. The signal $r \in \mathbb{R}^n$ is the trial invariant reference signal during the observation time interval, and $e_k = r - y_k$ the error signal. The state $x_k \in \mathbb{R}^p$ represents the time domain state x_{n_1} during trial k, and $x_0 = L_o r$ the state as function of external disturbance r. Finally, $u_k \in \mathbb{R}^p$ is the trial domain state vector.

As discussed in Section 3.1, P_H corresponds to the time-windowed system $W_o P W_a$. Moreover, $L_c \in \mathbb{R}^{m \times p}$ and $L_o \in \mathbb{R}^{p \times n}$ constitute the Hankel ILC controller with $\text{rank}(L_c) = \text{rank}(L_o) = p$, and w^{-1} represents the one trial domain backward shift operator: $u_k = w^{-1} u_{k+1}$.

Based on Figure 8, the trial domain dynamics of the Hankel ILC controlled system are given by

$$u_{k+1} = u_k + L_o(r - y_k), \quad f_k = L_c u_k, \quad u_0 = 0 \tag{13}$$

$$u_{k+1} = (I_p - L_o P_H L_c)u_k + L_o r. \tag{14}$$

These dynamics are used to study the convergence (stability) and performance properties of the Hankel ILC controlled system. The properties will subsequently be used to analyze the Hankel ILC controller design.

3.2.1 Convergence

Convergence of the Hankel ILC controlled system is essential to improve reading performance for cracked discs. With (14) describing the evolution of the state of the system in trial domain, the system in Figure 8 is convergent if and only if $\rho(I_p - L_oP_HL_c) < 1$, with $\rho(\cdot) = \max|\lambda_i(\cdot)|$ and $|\lambda_i|$ the absolute value of the ith eigenvalue.

If the system indeed is convergent, then it is guaranteed that the state u_k will converge to an asymptotic value $u_\infty = \lim_{k\to\infty} u_k$ after an infinite number of trials. It does, however, not provide any information about transient behaviour of f_k and y_k between $k = 0$ and $k \to \infty$. To avoid poor learning transients, Monotonic Convergence (MC) properties of the ILC controlled system are useful, see [10], [17].

Specifically, the following conditions are useful for analyzing MC:

$$\text{MC in } f_k: \quad \|T^{-1}(I_p - L_oP_HL_c)T\|_{i2} < 1, \tag{15}$$
$$\text{with } T \text{ such that } T^T L_c^T L_c T - I_p$$

$$\text{MC in } e_k: \quad \|T^{-1}(I_p - L_oP_HL_c)T\|_{i2} < 1, \tag{16}$$
$$\text{with } T \text{ such that } T^T L_c^T P_H^T P_H L_c T = I_p$$

in which $i2$ means induced 2-norm. In this chapter only MC in f_k is considered. For an extension to MC in e_k and convergence conditions for uncertain systems, the reader is referred to [17].

3.2.2 Performance

In ILC, performance refers to the capability of the ILC controller to minimize the asymptotic error $e_\infty = \lim_{k\to\infty} e_k$ in a certain sense. For the application in this chapter, high performance implies successful reading of information on a cracked disc, which is achieved if the focus and radial errors remain within certain predefined specifications.

Using (14) and the fact that $u_{k+1} = u_k$ for $k \to \infty$, e_∞ is given by

$$e_\infty = (I_n - P_HL_c(L_oP_HL_c)^{-1}L_o)r. \tag{17}$$

From (9), $x_0 - x_\infty = 0 \Rightarrow e_\infty = 0$. Using Figure 8

$$\Gamma(x_o - x_\infty) = L_o e_\infty \tag{18}$$

$$= L_o(I_n - P_H L_c (L_o P_H L_c)^{-1} L_o) r. \tag{19}$$

Clearly, provided that the inverse in (19) exists, $x_o - x_\infty = 0$, hence perfect tracking is achievable.

3.3 Hankel ILC control design

The Hankel ILC controller used for the cracked CD application is based on an inverse model ILC controller. As a result, L_o and L_c are given by

$$L_o = \Gamma P_o^\dagger, \quad L_c = P_c^\dagger + (I_m - P_c^\dagger P_c) M_c. \tag{20}$$

where $P_c^\dagger = P_c^T (P_c P_c^T)^{-1}$ is the Moore-Penrose inverse of P_c, $P_o^\dagger = (P_o^T P_o)^{-1} P_o^T$ is the Moore-Penrose inverse of P_o, Γ is a user defined learning matrix, and M_c is a filter that can be used to exploit non-uniqueness in the feedforward signal f_k. The reader is referred to [16] for further details on the design of M_c.

Convergence of the Hankel ILC controlled system with controller (20) is based on $I_p - L_o P_H L_c = I_p - \Gamma$. Based on the choice for Γ, the system can be made convergent. In fact, for $\Gamma = \gamma I$ with $\gamma \in (0, 2)$, it holds that the ILC controlled system is monotonically convergent in f_k. In addition, Γ can be used as a design parameter to ensure robust convergence of the ILC controlled system in the presence of model uncertainty. Finally, with L_o and L_c satisfying the rank condition, $x_\infty - x_0 = 0$ can be achieved.

A specific choice regarding the design of L_o and L_c is based on the singular value decomposition of P_H, see [16], which in fact amounts to a full rank decomposition of P_H. The singular value decomposition of P_H is defined by

$$P_H = \begin{bmatrix} U_1 & U_2 \end{bmatrix} \begin{bmatrix} \Sigma_1 & 0 \\ 0 & 0 \end{bmatrix} \begin{bmatrix} V_1^T \\ V_2^T \end{bmatrix} = U_1 \Sigma_1 V_1^T, \tag{21}$$

$$P_o = U_1, \quad P_c = \Sigma_1 V_1^T. \tag{22}$$

Moreover, let M_c be designed such that the energy of the weighted feedforward signal $\|f_\infty\|_W = f_\infty^T W f_\infty$ is minimized, with $W \in \mathbb{R}^{m \times m}$ a diagonal weighting matrix that separately penalizes every actuation sample in order to shape f_∞. Then L_o and L_c are given by

$$L_o = \Gamma U_1^T \tag{23}$$

$$L_c = (I_m - V_2 (V_2^T W V_2)^{-1} V_2^T W) V_1 \Sigma_1^{-1}. \tag{24}$$

In the next sections, L_o and L_c, as defined by the singular value decomposition in (23) and (24), are used to iteratively determine a command signal that enables the reading of a cracked optical disc.

4 Implementation aspects

In this section, certain implementation aspects that are inherently introduced by the specific application, which is the reading of cracked optical discs, are discussed. Firstly, trial-varying setpoint variations are discussed in Section 4.1. Then, in Section 4.2, an approach to deal with the DEFO, i.e., the defect detector, is presented. In Section 4.3, a state transformation is suggested that transforms the states introduced by the singular value decomposition into physical states being the position and velocity of the lens. Finally, the resulting Hankel ILC scheme is presented in Section 4.4.

4.1 Trial-varying setpoint variations

From physical considerations, the crack dimensions vary over the surface of the cracked disc. In virtue of Assumption 4 in Section 2.2, which is supported by Figure 5, the trial domain dynamics of the crack are significantly slower compared to the dynamics of the ILC controller, hence ILC can effectively attenuate disturbances caused by crack variations, see [5, Section 4.3.2]. In particular, in the setting of Figure 8, the trial-varying behaviour of the crack can be represented by the reference r. Although this signal r is unavailable in the implementation, since the error signal is directly reconstructed from reflected light, see also Section 2.1, the crack behaviour can conceptually be analyzed using trial variations in r.

4.2 Dealing with the DEFO

The DEFO switches off the updating of the feedback controller states in case an optical defect is detected to avoid excessive control inputs due to unreliable measurement data. For the application of Hankel ILC to cracked discs, this implies that the actuation of the lens during W_a involves an open-loop system, whereas the relevant tracking error during W_o is measured in a closed-loop situation. The reader is referred to Section 3 for the definition of W_o and W_a. Hence, in open-loop the dynamical behaviour is given by

$$e = r - P_{sys} f_k, \tag{25}$$

whereas in closed-loop the dynamical behaviour is given by

$$g_k = W_o \mathscr{S} (I + P_{sys} C)^{-1} (r - P_{sys} \mathscr{H} W_a f_k) \tag{26}$$

where $\mathscr{S}$ and $\mathscr{H}$ denote the ideal sampler and zero-order-hold interpolator, respectively. To reconstruct the error that would have resulted in open-loop, e_k is recon-

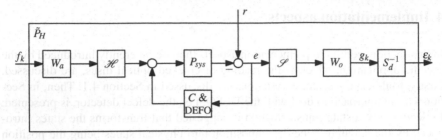

Fig. 9 Plant including open-loop error reconstruction.

structed by

$$\varepsilon_k = S_d^{-1} g_k, \tag{27}$$

where ε_k the reconstructed e_k and S_d^{-1} the inverse of the finite time convolution matrix representation of the discrete time LTI sensitivity function. The resulting plant including reconstruction of the open-loop error is denoted by $\tilde{P}_H$ and is depicted in Figure 9.

The discrete time sensitivity function S_d is obtained by computing the zero-order-hold equivalent of $S_c = (I + P_{sys}C)^{-1}$ based on the physical model in Figure 6, see also [1]. In Figure 10, Bode diagrams of the continuous time and discrete time inverse sensitivity functions are depicted. It is concluded that discretization errors are negligible. In case discretization errors are significant, the digital implementation aspects should be explicitly addressed to ensure the resulting ILC controller performs well, see also [12].

It should also be remarked that the inverse of the sensitivity may be unstable. Specifically, in Figure 10 the Bode diagrams of S_c^{-1} and S_d^{-1} are characterized by a high-gain low-frequent integrator, an underdamped resonance and a high-frequent asymptote of 0 [dB]. In this case, the integrator results in unstable behaviour. However, since the reconstruction is performed over a relatively short finite time interval n, the computation of the inverse is numerically reliable and the approach turns out to perform satisfactorily in practice.

4.3 State transformation to physical coordinates

The full rank decomposition of P_H based on an SVD does not necessarily result in a physically interpretable state $x_0 - x_k$. To gain more insight in the Hankel ILC control signals obtained after learning the cracked CDs, it is desirable to transform $x_0 - x_k$ to a state that can be physically interpreted. Specifically, this enables the identification of the crack geometry from learned command signals. Consider

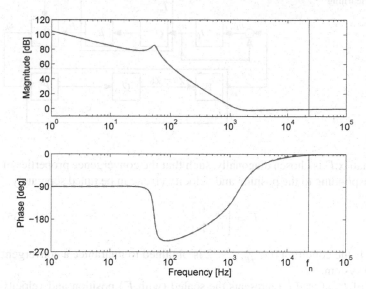

Fig. 10 Bode diagrams of S_c^{-1} (solid) and S_d^{-1} (dashed).

$$\chi_0 = \begin{bmatrix} \chi_p \\ \chi_v \end{bmatrix}, \tag{28}$$

with χ_p the focus or radial position error as result of the crack, and χ_v the physical velocity error. Then the goal is to determine a state transformation matrix Q such that

$$\chi_0 - \chi_k = Q(x_0 - x_k). \tag{29}$$

Using the fact that

$$\varepsilon_k = \underbrace{\begin{bmatrix} 1 & 0 \\ 1 & 1 \\ \vdots & \vdots \\ 1 & n-1 \end{bmatrix}}_{A} (\chi_0 - \chi_k), \tag{30}$$

Q is found by solving

$$\chi_0 - \chi_k = Q(x_0 - x_k) \tag{31}$$
$$= Q P_o^\dagger \varepsilon_k \tag{32}$$
$$= Q P_o^\dagger A(\chi_0 - \chi_k) \quad \Rightarrow \quad Q = (P_o^\dagger A)^{-1}. \tag{33}$$

As a consequence of the state transformation, the Hankel ILC filters L_o and L_c have to be altered to

$$\tilde{L}_o = \Gamma Q P_o^\dagger, \qquad \tilde{L}_c = L_c Q^{-1}. \tag{34}$$

Fig. 11 Applied learning loop.

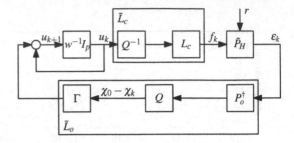

The learning matrix Γ is chosen diagonally, such that the convergence properties of the states corresponding to the position and velocity error can be tuned separately:

$$\Gamma = \begin{bmatrix} \gamma_p & 0 \\ 0 & \gamma_v \end{bmatrix} \tag{35}$$

In Section 3.2.1, the condition $0 < \gamma_p, \gamma_v < 2$ is obtained to guarantee a convergent ILC controlled system.

The output of $\tilde{L}_o$ at trial k represents the scaled (with Γ) position and velocity error between χ_0 and χ_k. Additional information about the crack properties can be obtained by investigating the converged state u_∞. For that purpose, it is assumed that $\tilde{P}_H \approx P_H$ and $\varepsilon_k \approx r - P_H f_k$. By design of L_o and L_c and based on the assumptions, the product $\tilde{L}_o \tilde{P}_H \tilde{L}_c \approx \Gamma$, and

$$u_\infty = (I - \tilde{L}_o \tilde{P}_H \tilde{L}_c) u_\infty + \tilde{L}_o r \tag{36}$$
$$= \Gamma^{-1} \chi_0. \tag{37}$$

Hence, after the Hankel ILC controlled system has converged, the trial domain state u_∞ equals the scaled focus or radial error position and velocity. This result is illustrated in Section 5. By defining A as in (30), the velocity state χ_v is scaled to [V/sample].

4.4 Resulting Hankel ILC scheme

Summarizing the implementation aspects in the previous sections, the DEFO and state transformation lead to the modified Hankel ILC controller depicted in Figure 11, where $\tilde{P}_H$ is depicted in Figure 9. This Hankel ILC controller is implemented in the next section to improve the reading of cracked optical discs.

5 Experimental results

In this section, the modified Hankel ILC controller in Figure 11 is implemented in the Philips BD1 player that is used for the recovery of data from a cracked disc. The measurements are performed with an effective actuation length of $\overline{m} = 24$, with $\overline{m} = 2m$ resulting in $m = 12$ individual sample values per state variable (χ_p or χ_v). The first measurement is performed in focus direction while reading the cracked CD-R, of which the geometry is given in Figure 5, at $x = 34$ [mm]. After convergence of the Hankel ILC algorithm, the measurement results in Figure 12 are obtained. In the considered cases, the Hankel ILC controller converged to a satisfactory command signal within 5 − 15 iterations.

In Figure 12 (top) the tracking error after convergence of the Hankel ILC controller is depicted, where the light colored (grey) interval indicates the defect interval that is detected by the DEFO. The tracking error signal in this light colored interval is unreliable and is neglected accordingly, i.e., it is neither used by the feedback controller nor by the Hankel ILC controller. The Hankel ILC controller results in a perfect compensation of the crack. For a comparison with the initial tracking error before the Hankel ILC controller is applied, the reader is referred to Figure 7.

In Figure 12 (middle), the required feedforward signal, which is equal to the actuator input, is depicted. Observe that in the defect interval, this signal almost saturates at 5.2 [V] indicating a suitable choice for $\overline{m}$. In Figure 12 (bottom), the reconstructed trajectory of the lens is shown, which is obtained by using the model P in (2). This trajectory shows a compensation in position (E_p) of 15 [μm] and in velocity (E_v) of -4[μm/ms] for which the latter corresponds to α=-1.7 [mrad]. This resembles the trajectory indicated by the arrow in Figure 5. Note that E_p is beyond the observation range of $\pm$ 2 [μm]. This implies that the Hankel ILC controller enables reading of cracked optical discs that were previously considered as unreadable.

A similar measurement result in radial direction is presented in Figure 13. The resulting trajectory shows that $E_p = 4.2$ [μm] and $E_v = -2.2$ [μm/ms] where the latter corresponds to γ=-1.0 [mrad]. As a result, the tracking error is compensated for outside the observation range of 0.8 [μm]. In the radial direction, it is important to check the continuity of the data track, i.e. the right track is being followed. The data channel provides this information. In the considered experiment, the correct track is followed after the crack, enabling perfect reconstruction of the data.

6 Conclusions

In this chapter, a novel ILC algorithm, referred to as Hankel ILC, has been further developed and implemented to enable the reading of cracked optical discs. In essence, the presented algorithm aims at perfectly steering the actuator during a crack towards the beginning of the track immediately *after* the crack has passed. Improved tracking properties of the resulting Hankel ILC controlled system in con-

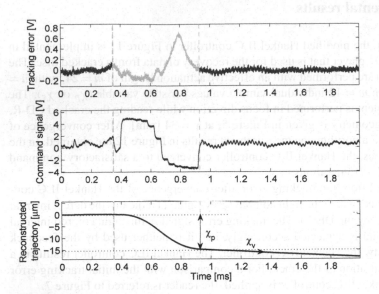

Fig. 12 Measurement results in focus direction.

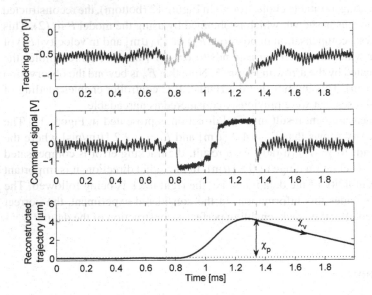

Fig. 13 Measurement results in radial direction.

junction with data redundancy in optical discs should enable the recovery of data from damaged discs that were otherwise considered unreadable.

Experimental results resulting from a commercial optical disc drive indeed confirm improved tracking properties of the Hankel ILC controlled system. Specifically, measurements show almost full compensation of the tracking errors in both focus and radial directions. These compensated discontinuities lie significantly outside the observation range of the optics. Hence, the proposed Hankel ILC controller can indeed enable the recovery of data from damaged discs that were previously considered as lost.

Open issues include robustness analysis of the approach with respect to variation in disc cracks and for a large number of discs an actual comparison of the recovered data compared to prior approaches, i.e., in case only the DEFO is used. This should also be used to quantify the qualitative properties of the considered class of cracked discs in Section 2.2. In addition, the Hankel ILC controller can be used to recover data due to related disc defects, see [8] for an overview of relevant defects. In addition, implementation aspects, both regarding the DEFO and state reconstruction in Section 4.2 and Section 4.3, respectively, should be further theoretically justified, as well as the theoretical aspects associated with the switching dynamics of the overall system.

References

[1] Åström, K.J., Wittenmark, B.: Computed-Controlled Systems: Theory and Design, second edn. Prentice-Hall, Englewood Cliffs, NJ, USA (1990)

[2] Bien, Z., Xu, J.X.: Iterative Learning Control: Analysis, Design, Integration and Applications. Kluwer Academic Publishers, Norwell, MA, USA (1998)

[3] Bristow, D.A., Tharayil, M., Alleyne, A.G.: A survey of iterative learning control: A learning-based method for high-performance tracking control. IEEE Control Systems Magazine 26(3), 96–114 (2006)

[4] Coastor: Has anyone tried burning a disc with a cracked inner ring? http://club.cdfreaks.com (2007)

[5] Dijkstra, B.: Iterative learning control, with applications to a wafer-stage. Ph.D. thesis, Delft University of Technology, Delft, The Netherlands (2004)

[6] Dijkstra, B.G., Bosgra, O.H.: Extrapolation of optimal lifted system ILC solution, with application to a waferstage. In: Proceedings of the 2002 American Control Conference, pp. 2595–2600. Anchorage, AK, USA (2002)

[7] Gorinevsky, D.: Loop shaping for iterative control of batch processes. IEEE Control Systems Magazine 22(6), 55–65 (2002)

[8] Van Helvoirt, J., Steinbuch, M., Leenknegt, G.A.L., Goossens, H.J.: Disc defect classification for optical disc drives. IEEE Transactions on Consumer Electronics 51(3), 856–863 (2005)

[9] Jutte, P.T.: Light path design for optical disk systems. Ph.D. thesis, Eindhoven University of Technology, Eindhoven, The Netherlands (2005)

[10] Longman, R.W.: Iterative learning control and repetitive control for engineering practice. International Journal of Control 73(10), 930–954 (2000)

[11] Moore, K.L.: Iterative learning control - an expository overview. Applied and Computational Controls, Signal Processing, and Circuits 1, 151–214 (1999)

[12] Oomen, T., van de Wijdeven, J., Bosgra, O.: Suppressing intersample behavior in iterative learning control. Automatica 45(4), 981–988 (2009)

[13] Pintelon, R., Schoukens, J.: System Identification: A Frequency Domain Approach. IEEE Press, New York, NY, USA (2001)

[14] Starwing, F.: Disappointed: Cracked disc. http://utforums.epicgames.com (2007)

[15] Van de Wijdeven, J.: Iterative learning control design for uncertain and time-windowed systems. Ph.D. thesis, Eindhoven University of Technology, Eindhoven, The Netherlands (2008)
[16] Van de Wijdeven, J., Bosgra, O.: Residual vibration suppression using Hankel iterative learning control. International Journal of Robust and Nonlinear Control 18(10), 1034–1051 (2008)
[17] Van de Wijdeven, J., Donkers, T., Bosgra, O.: Iterative learning control for uncertain systems: Robust monotonic convergence analysis. Submitted for publication (2009)
[18] Zhou, K., Doyle, J.C., Glover, K.: Robust and Optimal Control. Prentice Hall, Upper Saddle River, NJ, USA (1996)

Advances in Data-driven Optimization of Parametric and Non-parametric Feedforward Control Designs with Industrial Applications

Rob Tousain and Stan van der Meulen

Abstract The performance of many industrial control systems is determined to a large extent by the quality of both setpoint and disturbance feedforward signals. The quality that is required for a high tracking performance is generally not achieved when the controller parameters are determined on the basis of a detailed model of the plant dynamics or manual tuning. This chapter shows that the optimization of the controller parameters by iterative trials, i.e., data-driven, in both parametric and non-parametric feedforward control structures avoids the need for a detailed model of the plant dynamics, achieves optimal controller parameter values, and allows for the adaptation to possible variations in the plant dynamics. Two industrial applications highlight the large benefits of the data-driven optimization approach. The optimization of the feedforward controller parameters in a wafer scanner application leads to extremely short settling times and higher productivity. The optimization of the current amplifier setpoints in a digital light projection (DLP) application leads to nearly constant color rendering performances of the projection system in spite of large changes in the lamp dynamics over its life span.

1 Introduction

Performance requirements in industrial high-precision electromechanical servo systems, such as pick-and-place robots, printed circuit board (PCB) assembly robots, laser welding robots, printers, and wafer scanners are ever increasing. Accuracy requirements tend to tighten due to the trend of miniaturization in the semiconductor industry and the manufacturing industry in general. To reduce manufacturing

Rob Tousain
Philips Lighting, Mathildelaan 1, 5611 BD, Eindhoven e-mail: Rob.Tousain@philips.com

Stan van der Meulen
Eindhoven University of Technology, Department of Mechanical Engineering, Control Systems Technology Group, PO Box 513, 5600 MB, Eindhoven e-mail: S.H.v.d.Meulen@tue.nl

P.M.J. Van den Hof et al. (eds.), *Model-Based Control: Bridging Rigorous Theory and Advanced Technology*, DOI: 10.1007/978-1-4419-0895-7_10,
© Springer Science + Business Media, LLC 2009

costs and cycle times, productivity has to rise, leading to very aggressive setpoints. Illustratively, the setpoint acceleration of present-day wafer scanners achieves levels of 70 $[m/s^2]$ at wafer level, compared to levels of 10 $[m/s^2]$ about ten years ago. To achieve the performance requirements, industrial motion control systems are equipped with feedback controllers and feedforward controllers. The feedback controller ensures stability and improves disturbance rejection [21, 28], whereas the feedforward controller improves tracking performance [15]. The design of the feedforward controller is no less crucial to meeting the performance requirements than the design of the feedback controller. Various approaches to designing the feedforward controller are known in literature.

In *parametric feedforward control*, i.e., model-based feedforward control, the feedforward controller is equal to the inverse of the model of the plant. Various examples of this approach can be found in [23, 13, 8, 24, 3]. In this approach, the quality of the feedforward signal is limited by the quality of the model of the plant. Often, simplified versions of the model-inversion approach are implemented. A well-known strategy is to use a linear combination of the setpoint acceleration and the setpoint velocity, see, e.g., [15, 4]. This type of feedforward control is simple and can be highly effective, where the acceleration-related part of the feedforward signal compensates for inertia forces and the velocity-related part of the feedforward signal for damping forces. The controller parameters of this feedforward controller are mostly obtained by manual tuning, which limits the quality of the feedforward signal. Adaptive feedforward control can be used to tune and adapt the feedforward controller parameters to measurement data in order to accommodate for changes in the plant dynamics, see, e.g., [12, 27, 29]. However, adaptive feedforward control is less suited for the application to finite time tasks, due to the adaptation at each sample instant. Also, adaptive feedforward control requires the condition of persistence of excitation to be satisfied, which may impose undesired requirements on the setpoint.

In *non-parametric feedforward control*, the feedforward signal is equal to a time series, which is directly manipulated and implemented. This type of feedforward control is typically used in cases where the same setpoint is continuously executed and where ultimate tracking performance is required. In such cases, a dedicated feedforward time series is defined for the specific setpoint that is executed. The maximum design freedom that is available in the time series can then be used to eliminate the deterministic error patterns completely. The optimization of such a dedicated feedforward time series can be done by using iterative learning control (ILC). Essentially, this technique determines the feedforward signal by iterative trials, where convergence of the update law is generally determined by a model of the plant. Excellent overviews on the subject of ILC can be found in [18, 16, 5, 2] and pioneering work on the application of ILC in a wafer scanner can be found in [6].

The utilization of iterative trials in ILC has attractive properties, since it avoids the need for detailed knowledge of the plant and it allows for the normal operation of the plant. Actually, ILC is a specific direct tuning method, in which a controller parameter optimization problem is formulated and the basic idea is to use numerical optimization and to use measurement data from iterative trials in order to optimize the controller parameters. In ILC, the controller parameters are represented by the individual samples of the feedforward signal. In order to reduce the number of controller parameters, the introduction of basis functions is possible, see, e.g., [10]. Obviously, the quality of the feedforward signal depends on the selection of these basis functions, which is mainly determined by the specific application and is often not straightforward in practice.

The main contribution of this chapter concerns the selection of parametric and non-parametric feedforward control structures and the optimization of the controller parameters by iterative trials for two industrial applications. This data-driven optimization approach avoids the need for a detailed model of the plant dynamics, achieves optimal controller parameter values, and allows for the adaptation to possible variations in the plant dynamics, which is generally not achieved by approaches in which the controller parameters are determined on the basis of a detailed model of the plant dynamics or manual tuning. The organization of this chapter is as follows. The direct tuning method that is used to optimize the controller parameters in the feedforward controller by iterative trials is discussed in Section 2. In Section 3, the selection of a parametric feedforward control structure for a wafer stage application is considered, where experimental results are shown to demonstrate the benefits of the iterative optimization procedure. In Section 4, the selection of a non-parametric feedforward control structure for a digital light projection (DLP) application is considered, where the relation between the direct tuning method in Section 2 and ILC is highlighted. Again, experimental results are shown to demonstrate the benefits of the iterative optimization procedure. Finally, conclusions are drawn in Section 5.

2 Data-driven Feedforward Control Optimization

Consider the one-degree-of-freedom control architecture in Fig. 1. Here, P denotes the plant and K_{fb} denotes the feedback controller, which are discrete time, single input single output (SISO), and linear time-invariant (LTI). Furthermore, the setpoint is denoted by r, the tracking error by e, the feedback signal by u_{fb}, the feedforward signal by u_{ff}, the plant input by u, the disturbances by w, and the plant output by y. When the setpoint is equal to a finite time task, i.e., a trial, the values of each discrete time signal $x(k)$ in trial l are collected in a time series x^l, which is defined by:

$$x^l = [\, x(0) \;\; \cdots \;\; x(N-1) \,]^T, \tag{1}$$

for $k = 0, \ldots, N-1$. Here, k denotes the sample instant and N denotes the number of samples in trial l. The feedforward signal u_{ff}^l is defined by:

$$u_{ff}^l = \xi^l \theta^l, \tag{2}$$

where ξ^l has size $N \times n$ and θ^l has size $n \times 1$. Here, n denotes the number of controller parameters in the controller parameter vector θ^l. The selection of both ξ^l

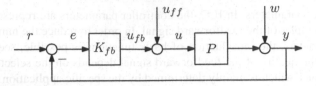

Fig. 1 System with a one-degree-of-freedom control architecture.

and θ^l is highlighted in Sections 3 and 4. To achieve a high tracking performance, the quality of the feedforward signal u^l_{ff} is of crucial importance. This quality is obtained by means of a direct tuning method, in which the basic idea is to use numerical optimization and to use measurement data from iterative trials in order to optimize the controller parameter vector θ^l. In this way, optimal controller parameter values are obtained for the actual plant, which is generally not achieved by, e.g., model-based feedforward control and manual tuning.

2.1 Objective Function

The direct tuning optimization problem is defined by:

$$\min_{\theta^l} V(\theta^l), \tag{3}$$

where V denotes the objective function that represents the control system performance. To obtain a high tracking accuracy, the signal-based objective function V is chosen equal to the square of the 2-norm of the tracking error:

$$V(\theta^l) = e^{l^T}(\theta^l)e^l(\theta^l). \tag{4}$$

2.2 Optimization Algorithm

Constraints on the controller parameters are neglected, i.e., the direct tuning optimization problem is unconstrained, although this is not essential. A well-known optimization algorithm for unconstrained optimization is given by Newton's method. Newton's method is given by the following update law for the controller parameter vector θ^l in trial l of the iterative optimization procedure:

$$\theta^{l+1} = \theta^l - \alpha^l \left(\nabla^2 V(\theta^l) \right)^{-1} \nabla V(\theta^l), \tag{5}$$

see [19]. Here, α is the step length, ∇V is the gradient of the objective function, and $\nabla^2 V$ is the Hessian of the objective function. The gradient of the objective function

Eq. (4) with respect to the controller parameters is given by:

$$\nabla V(\theta^l) = 2\nabla e^{l^T}(\theta^l)e^l(\theta^l),\tag{6}$$

whereas the Hessian of the objective function Eq. (4) with respect to the controller parameters is given by:

$$\nabla^2 V(\theta^l) = 2\nabla e^{l^T}(\theta^l)\nabla e^l(\theta^l).\tag{7}$$

The application of Newton's method to the direct tuning optimization problem leads to the direct tuning method described below.

Direct tuning method

1) Set the trial number l equal to $l = 0$.
2) Set the initial controller parameter values θ^0.
3) Execute a finite time task r^l and measure the tracking error e^l.
4) Evaluate the objective function:

$$V(\theta^l) = e^{l^T}(\theta^l)e^l(\theta^l).\tag{8}$$

Proceed with Step 5 if the objective function value is not satisfactory. Otherwise, proceed with Step 6.
5) Execute the optimization algorithm:

$$\theta^{l+1} = \theta^l - \alpha^l \left(\nabla^2 V(\theta^l)\right)^{-1}\nabla V(\theta^l).\tag{9}$$

6) Set the trial number l equal to $l = l + 1$. Proceed with Step 3.

Several ways are known to approximate ∇V and $\nabla^2 V$, see [17]. Approximations are inevitable, since the actual plant is unknown. Here, ∇V and $\nabla^2 V$ are approximated by using both model knowledge and measurement data. This approach to finding the gradient and the Hessian is based on lifting the system executing repeating tasks from *time* domain to *trial* domain, see, e.g., [20, 25]. This is explained below. Consider the following state-space representation of a discrete time, SISO, LTI system:

$$
\begin{align}
x(k+1) &= Ax(k) + Bu(k) \tag{10a}\\
y(k) &= Cx(k) + Du(k), \tag{10b}
\end{align}
$$

where $u(k)$ denotes the input and $y(k)$ denotes the output. The lifted system description of this system executing repeating tasks of length N is defined by:

$$
y^l = \underbrace{\begin{bmatrix} D & 0 & \cdots & \cdots & 0 \\ CB & D & \ddots & & \vdots \\ \vdots & CB & \ddots & \ddots & \vdots \\ \vdots & \vdots & & \ddots & 0 \\ CA^{N-2}B & CA^{N-3}B & \cdots & CB & D \end{bmatrix}}_{\mathscr{T}_{u^l \to y^l}} u^l, \tag{11}
$$

for $k = 0, \ldots, N - 1$. Here, the $N \times N$ Toeplitz matrix $\mathscr{T}_{u^l \to y^l}$ contains the first N Markov parameters of the discrete time system. For the one-degree-of-freedom control architecture in Fig. 1, the map between the position setpoint r^l and the tracking error e^l is defined by the sensitivity Toeplitz matrix $\mathscr{T}_{r^l \to e^l}$, whereas the map between the feedforward signal u^l_{ff} and the plant output y^l is defined by the process sensitivity Toeplitz matrix $\mathscr{T}_{u^l_{ff} \to y^l}$.

With these definitions, the tracking error in the trial domain is defined by:

$$
e^l(\theta^l) = \overline{\mathscr{T}_{r^l \to e^l}}(r^l - w^l) - \overline{\mathscr{T}_{u^l_{ff} \to y^l}} \xi^l \theta^l, \tag{12}
$$

where the overbar indicates that the corresponding Toeplitz matrix describes the actual plant. Using model approximations $\mathscr{T}_{r^l \to e^l}$ and $\mathscr{T}_{u^l_{ff} \to y^l}$, this relation can be approximated as follows:

$$
e^l(\theta^l) = \mathscr{T}_{r^l \to e^l}(r^l - w^l) - \mathscr{T}_{u^l_{ff} \to y^l} \xi^l \theta^l. \tag{13}
$$

Using Eq. (13), it is possible to derive the gradient of the error signal with respect to the controller parameters:

$$
\nabla e^l(\theta^l) = -\mathscr{T}_{u^l_{ff} \to y^l} \xi^l. \tag{14}
$$

The gradient and the Hessian are obtained by substituting Eq. (14) into Eq. (6) and Eq. (7), respectively.

2.2.1 Convergence

Convergence to a (local) minimum of the objective function is guaranteed if Newton's method is combined with line search optimization, see [7]. When a constant value for the step length is employed, convergence can be proven differently. Substitution of Eq. (6) and Eq. (7) into Eq. (5) and using Eq. (12) and Eq. (14), leads to a linear discrete time system:

$$\theta^{l+1} = \underbrace{\left(I - \alpha^l \left(\xi^{l^T} \mathscr{T}_{u_{ff}^l \to y^l}^T \mathscr{T}_{u_{ff}^l \to y^l} \xi^l\right)^{-1} \xi^{l^T} \mathscr{T}_{u_{ff}^l \to y^l}^T \overline{\mathscr{T}_{u_{ff}^l \to y^l}}^l \xi^l\right)}_{A} \theta^l +$$

$$+ \underbrace{\alpha^l \left(\xi^{l^T} \mathscr{T}_{u_{ff}^l \to y^l}^T \mathscr{T}_{u_{ff}^l \to y^l} \xi^l\right)^{-1} \xi^{l^T} \mathscr{T}_{u_{ff}^l \to y^l}^T \overline{\mathscr{T}_{r^l \to e^l}}^l (r^l - w^l)}_{B}, \quad (15)$$

where I denotes the unit matrix of appropriate dimensions. From linear system theory, it is well-known that a linear discrete time system is stable if the eigenvalues λ of the state matrix A are all within the unit circle, i.e., $|\lambda(A)| < 1$ for all λ, see [14]. From Eq. (15), it follows that the eigenvalues λ of the state matrix A are equal to $1 - \alpha^l$ if the approximated process sensitivity Toeplitz matrix is equal to the actual process sensitivity Toeplitz matrix.

2.2.2 Implementation

The direct tuning method can be used for several purposes. It can be used as an initial automatic tuning of the control system. However, if the plant suffers from position-, load state-, or time-dependency, it may be beneficial to have the iterative optimization procedure update the controller parameters continuously. This way, the feedforward controller can be made to adapt to such changes of the plant. Since the dynamics that describe the parameter variations from trial to trial is given by a discrete time LTI system (Eq. (15)), the adaptation performance of the iterative optimization procedure can be analyzed using linear system theory, see [17].

3 Parametric Feedforward Control Optimization for a Wafer Stage Application

The feedforward control optimization procedure from Section 2 is applied to a high-precision wafer stage that is part of a wafer scanner. A wafer scanner is used in the mass production process of integrated circuits (ICs), see [22], where it is responsible for the photo lithographic process in which the IC pattern is printed onto a silicon disk, i.e., a wafer. In a wafer scanner, the wafer stage is the high-precision electromechanical servo system that positions the wafer with respect to the imaging optics. As a result, the wafer stage determines the throughput and the quality of the products to a large extent and it is subject to severe performance requirements. Typical velocities and accelerations are 0.5 [m/s] and 50 [m/s²], respectively, whereas the tracking accuracy is in terms of nanometers and micro radians. The exposures of the IC patterns onto the wafer are rapidly started after accelerating the wafer stage to constant velocity, when the tracking errors have settled within tight boundaries.

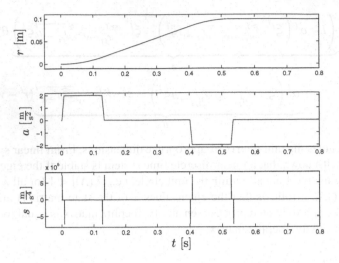

Fig. 2 The position r, the acceleration a, and the snap s of the point-to-point motion.

A shorter settling time leads to a higher wafer scanner productivity. Hence, optimal controller parameter values are of crucial importance.

The wafer stage is actuated and controlled in six degrees of freedom: three translations (x, y, and z) and three rotations (R_x, R_y, and R_z, where the subscripts refer to the rotation axis). Here, the wafer stage dynamics in the y-direction is considered, which is the main scan direction. A typical finite time task that is executed in the y-direction is given by a point-to-point motion, see Fig. 2. The frequency response function (FRF) measurements and a very simple second-order discrete time transfer function model are depicted in Fig. 3. The Bode diagram of the discretized feedback controller is depicted in Fig. 4. Although the second-order discrete time transfer function model from Fig. 3 is not very detailed, it is sufficiently accurate for the iterative optimization procedure to be convergent.

3.1 Feedforward Controller Parameterization

For the wafer stage, the feedforward signal u_{ff} is composed of two basis functions, i.e., a parametric feedforward controller, which implies that $n = 2$. The basis functions are given by the fourth- and second-order derivative of the position setpoint (s and a, respectively), with weighting factors kfs and kfa. The signals s and a are typically referred to as the *snap* setpoint and the *acceleration* setpoint, respectively. The motivation for this parametric feedforward controller is found in [4], which shows that acceleration feedforward exactly compensates for the rigid body mode, whereas snap feedforward exactly compensates for the low-frequency contributions of all residual plant modes. For a general motion system, two additional

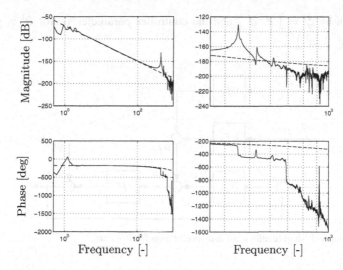

Fig. 3 Bode diagram of the wafer stage dynamics in the y-direction, where the figures on the right are close-ups of the figures on the left (solid lines: frequency response function measurements; dashed lines: second-order discrete time transfer function model) [17].

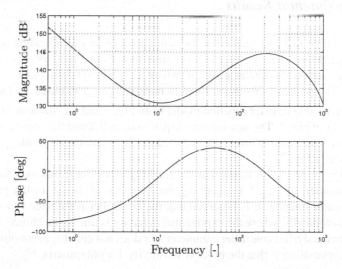

Fig. 4 Bode diagram of the discretized feedback controller for the wafer stage dynamics in the y-direction [17].

basis functions are given by the third- and first-order derivative of the position setpoint (j and v, respectively), with weighting factors kfj and kfv. The signals j and v are typically referred to as the *jerk* setpoint and the *velocity* setpoint, respectively. More details can be found in [15]. The control system that is obtained with this parametric feedforward controller is depicted in Fig. 5. As a result, the matrix

ξ^l is given by $\xi^l = [\ s^l\ \ a^l\ ]$ and the controller parameter vector θ^l is given by $\theta^l = [\ kfs^l\ \ kfa^l\ ]^T$.

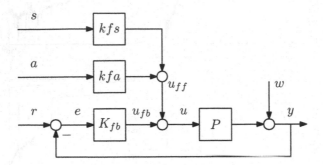

Fig. 5 System with snap feedforward and acceleration feedforward.

3.2 Experimental Results

Each finite time task is executed at the same position in the operating area of the wafer stage. More specifically, the finite time task from Fig. 2 is executed in the y-direction, across the center position, where $y(0) = -0.05$ [m] and $y(N-1) = 0.05$ [m]. The process sensitivity Toeplitz matrix $\mathscr{T}_{u^l_{ff} \to y^l}$ is based on the second-order discrete time transfer function model from Fig. 3 and the discretized feedback controller from Fig. 4. The step length is equal to $\alpha^l = 0.8$ and the number of trials is equal to six. The tracking errors are depicted in Fig. 6 and the controller parameters are depicted in Fig. 7.

From Fig. 6, it is concluded that the tracking error decreases as a function of the trial number. In Fig. 7, the convergence behavior of the controller parameters is of particular interest. It is observed that the controller parameter kfa does converge monotonically, where the exponential behavior is due to the choice of the step length. Obviously, the controller parameter kfs does not converge monotonically. A possible explanation is that the process sensitivity Toeplitz matrix $\mathscr{T}_{u^l_{ff} \to y^l}$ is based on the second-order discrete time transfer function model from Fig. 3, where the resonant dynamics is not taken into account.

4 Non-parametric Feedforward Control Optimization for a Digital Light Projection Application

The application of ultra high pressure (UHP) lamps in digital light projection (DLP) systems sets new and more stringent requirements on the quality of the light gen-

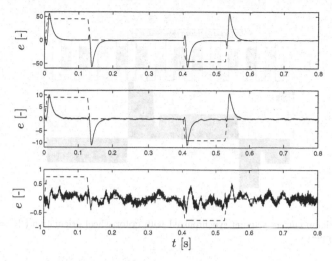

Fig. 6 Experimental tracking errors in trial 0 (top), 1 (middle), and 5 (bottom) (solid lines: tracking error; dashed lines: scaled acceleration setpoint) [17].

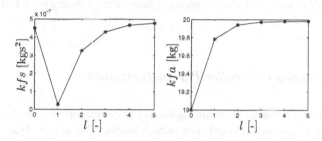

Fig. 7 Experimental controller parameters.

erated by these lamps. Developments like dark video enhancement (DVE) for DLP systems require besides stability an accurate light-output that changes very quickly with different segments of the color wheel. In Fig. 8, a typical relation between the desired lamp current (and related light) and color wheel position is given. To avoid picture artifacts, the light and hence the lamp current must track these setpoints very accurately. Up to now, this has been realized by means of empirically obtained tables. In this case, the lamp driver current setpoints are modified according to these tables, in such a way that satisfactory lamp performance is achieved over the lamp's life span.

There are several reasons why this strategy is no longer considered to be attractive. First, the manual tuning approach does not result in optimal behavior. Second, the manual tuning approach is very difficult and time-consuming, mainly because, to cope with changes in the lamp's behavior in time, different tables are required throughout the lamp's life span. Third, the manual tuning approach is rather inflexible with regard to changes in the lamp current setpoint. Although the DVE current

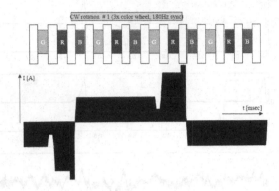

Fig. 8 UHP lamp current setpoint synchronized with color wheel orientation ('R' = red, 'G' = green, 'B' = blue).

setpoint is the latest development, new, even more exotic current setpoints are expected to follow soon. For these reasons, a new control strategy for UHP lamps is desired.

4.1 Feedforward Controller Parameterization

For the UHP lamp, the feedforward signal u_{ff} is composed of a time series, i.e., a non-parametric feedforward controller, which implies that $n = N$. Hence, the individual samples of the feedforward signal are the *controller parameters*. Here, the length of the feedforward signal and the setpoint are equal. However, a restriction of the length of the feedforward signal to a certain interval is also possible. As a result, the matrix ξ^l is given by $\xi^l = I$ and the controller parameter vector θ^l is given by $\theta^l = u_{ff}^l$. This non-parametric feedforward controller in combination with the direct tuning method, see Section 2, leads to a recursive update law that is very close to the lifted approach in iterative learning control (ILC).

4.2 Non-parametric Feedforward Control Optimization and ILC

ILC is often employed for the optimization of feedforward signals for systems that execute the same trajectory over and over again. Roughly, two classes of design approaches to ILC can be distinguished. The classic approach is based on frequency domain design, where use is made of a learning filter L and a robustness filter Q, see [1] and [18]. The lifted approach is based on time domain design, where use is made of a lifted representation of the learning control system, see [25] and [9].

A typical update law in lifted ILC is given by:

$$u_{ff}^{l+1} = u_{ff}^l + Le^l, \tag{16}$$

where L denotes the learning matrix. A more detailed motivation for the update law Eq. (16) can be found in [25, Section 3] and [9, Section 3.2.2]. A familiar expression for the learning matrix L is given by the regularized pseudoinverse:

$$L = \left(\mathscr{T}_{u_{ff}^l \to y^l}^T \mathscr{T}_{u_{ff}^l \to y^l} + \beta I \right)^{-1} \mathscr{T}_{u_{ff}^l \to y^l}^T, \tag{17}$$

see [11]. Here, β denotes a scaling factor, where it holds that $\beta \neq 0$. The scaling factor β is necessary to allow the computation of the inverse in Eq. (17) in case the process sensitivity is strictly proper, i.e., it does not have a constant (direct feed through) term. Besides the scaling factor β, the combination of the update law Eq. (16) and the learning matrix Eq. (17) is equal to the combination of the update law Eq. (5), the gradient of the objective function Eq. (6), the Hessian of the objective function Eq. (7), and the gradient of the error signal with respect to the controller parameters Eq. (14), where $\alpha^l = 1.0$, i.e., a full Newton step. Hence, if the objective function is chosen as a quadratic function of the tracking error, it is observed that each lifted ILC update is equal to a full Newton step, where the Hessian is fixed and the gradient is updated by using the measured tracking error from the previous trial.

4.3 ILC Design for UHP Lamp Current Control

Time-varying Dynamics The dynamics of UHP lamps are known to vary significantly in time, see [26]. Experimental pulse response modeling of the transfer from current setpoint to actual lamp current is done to generate accurate models to be used in the ILC design. The resulting measured pulse responses of a representative selection of two different lamps are shown in Fig. 9, where the solid lines represent the measured pulse responses and the dotted lines (shifted up by 0.2 [-] to make them distinguishable from the solid lines) the responses from a physical model, for details, see [26].

Window Selection In lifted ILC, the control and observation window can be selected individually. To make implementation of the ILC in inexpensive microprocessors feasible, computation and memory requirements should be limited to the minimum. Therefore, very short observation windows (25 samples only) and even shorter control windows (15 samples only) are selected. For the DVE platform, two control and two observation intervals are selected, as depicted in Fig. 10. The first control interval is located around the 'dip' in the current setpoint at the end of the green section, the second around the commutation of the current (switching from positive to negative or *vice versa*).

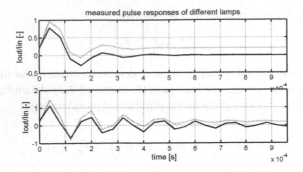

Fig. 9 Pulse responses of two different lamps measured at a sampling frequency of 30 [kHz]: a lamp with 3000 burning-hours (top) and a new lamp (bottom).

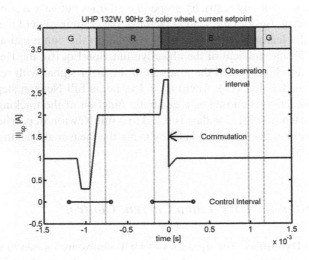

Fig. 10 Typical DVE lamp current setpoint with control and observation intervals.

4.4 Experimental Results

The ILC design has been tested together with various lamp/driver combinations. The results for two different 132 [W] lamps are shown: a brand new lamp, with a lamp voltage of 57 [V] and an old lamp (3000 burning-hours), with a lamp voltage of 117 [V]. The two lamps mark the extremes of the rather wide range of variations in lamp dynamics. The lamp current setpoints of the 57 [V] lamp without and with ILC are plotted in Fig. 11. With ILC, excellent tracking of the lamp current setpoint is obtained. Spectacular improvements are visible after commutation: the resonant response completely disappears. The dotted lines indicate the spokes of the color wheel. The lamp current setpoints of the 117 [V] lamp without and with ILC are plotted in Fig. 12. Observe the large difference in the response to the 'dip' between the 57 [V] lamp (Fig. 11) and the 117 [V] lamp (Fig. 12). Nevertheless, ILC manages

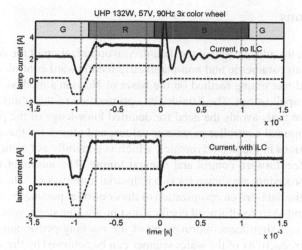

Fig. 11 Lamp current without (top) and with (bottom) ILC for 57 [V] lamp. Setpoint shifted down by -2 [A] and shown by dashed line.

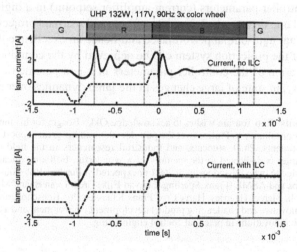

Fig. 12 Lamp current without (top) and with (bottom) ILC for 117 [V] lamp. Setpoint shifted down by -2 [A] and shown by dashed line.

to achieve excellent tracking of the lamp current setpoint, also for the 117 [V] lamp. Based on the successful results from the test setup, it was decided to make ILC part of the newest UHP lamp driver technology, called VIDI technology. Drivers with VIDI technology have been introduced in the market from March 2006 on. Over 1.5 million pieces have been sold in February 2009.

5 Conclusions

In this chapter, the added value of the data-driven optimization of the controller parameters in both parametric and non-parametric feedforward control structures by means of the direct tuning method on the basis of Newton's method is shown for two industrial applications. The data-driven optimization of the controller parameters by iterative trials avoids the need for detailed knowledge of the plant dynamics, achieves optimal controller parameter values, and allows for the adaptation to possible variations in the plant dynamics, which is generally not achieved by, e.g., model-based feedforward control and manual tuning. The merits of this iterative optimization procedure are shown for two industrial applications. The first application concerns the data-driven optimization of the controller parameters (acceleration feedforward and snap feedforward coefficients) in a wafer stage, which is used in wafer scanners. A significant improvement of the tracking performance and hence an improved productivity of the wafer scanner can be achieved by the simultaneous and data-driven optimization of the controller parameters in a parametric feedforward control structure. The second application concerns the data-driven optimization of the controller parameters (current amplifier setpoint) in a digital light projection (DLP) lamp driver, which is used in front- and rear-end projection systems. A nearly constant light-output performance and hence an excellent color rendering performance of the projection system can be achieved by the continuous and data-driven optimization of the controller parameters in a non-parametric feedforward control structure, in spite of large changes in the lamp dynamics over its life span.

Acknowledgements Rob Tousain wishes to acknowledge Okko Bosgra for his much-appreciated and very fruitful consultancy to Philips over the past decades and his great support and inspiration to many M.Sc. students, Ph.D. students, and industrial researchers in the field of systems and control. This chapter is dedicated to the memory of a wonderful, brilliant colleague and friend, Eduard van der Meché, who passed away in 2007 unexpectedly. Further, Rob Tousain owes much gratitude to Philips and ASML (Frank Sperling, Jan van Eijk, Gregor van Baars, Marc van de Wal, Maarten Steinbuch, Marcel Heertjes, Harry Cox, Frans Klaassen, Jos Benschop, and many others) for creating an innovative and challenging product development environment and a cradle for many advances and new innovations in practical control engineering.

References

[1] Arimoto, S.: Mathematical theory of learning with applications to robot control. In: K.S. Narendra (ed.) Adaptive and Learning Systems: Theory and Applications, pp. 379–388. Plenum Press, New York (1986)
[2] Bien, Z., Xu, J.X.: Iterative Learning Control – Analysis, Design, Integration, and Applications. Kluwer Academic Publishers, Boston (1998)
[3] Boerlage, M., Steinbuch, M., Lambrechts, P., Van de Wal, M.: Model-Based Feedforward for Motion Systems. In: Proceedings of the 2003 IEEE Conference on Control Applications, vol. 2, pp. 1158–1163. Istanbul, Turkey (2003)
[4] Boerlage, M., Tousain, R., Steinbuch, M.: Jerk Derivative Feedforward Control for Motion Systems. In: Proceedings of the 2004 American Control Conference, vol. 5, pp. 4843–4848. Boston, Massachusetts (2004)

[5] Bristow, D.A., Tharayil, M., Alleyne, A.G.: A Survey of Iterative Learning Control – A Learning-Based Method for High-Performance Tracking Control. IEEE Control Systems Magazine 26(3), 96–114 (2006)

[6] De Roover, D.: Motion Control of a Wafer Stage: A Design Approach for Speeding Up IC Production. Ph.D. Thesis, Delft University of Technology, Delft, The Netherlands (1997)

[7] Dennis Jr., J.E., Schnabel, R.B.: Numerical Methods for Unconstrained Optimization and Nonlinear Equations. No. 16 in Classics in Applied Mathematics. Society for Industrial and Applied Mathematics (SIAM), Philadelphia, Pennsylvania (1996)

[8] Devasia, S.: Should Model-Based Inverse Inputs be Used as Feedforward Under Plant Uncertainty? IEEE Transactions on Automatic Control 47(11), 1865–1871 (2002)

[9] Dijkstra, B.G.: Iterative Learning Control, With Applications to a Wafer-Stage. Ph.D. Thesis, Delft University of Technology, Delft, The Netherlands (2004)

[10] Frueh, J.A., Phan, M.Q.: Linear Quadratic Optimal Learning Control (LQL). International Journal of Control 73(10), 832–839 (2000)

[11] Ghosh, J., Paden, B.: A pseudoinverse-based iterative learning control. IEEE Transactions on Automatic Control 47(5), 831–837 (2002)

[12] Hägglund, T., Åström, K.J.: Industrial Adaptive Controllers Based on Frequency Response Techniques. Automatica 27(4), 599–609 (1991)

[13] Hunt, L.R., Meyer, G., Su, R.: Noncausal Inverses for Linear Systems. IEEE Transactions on Automatic Control 41(4), 608–611 (1996)

[14] Kailath, T.: Linear Systems. Prentice-Hall Information and System Sciences Series. Prentice-Hall, Englewood Cliffs, New Jersey (1980)

[15] Lambrechts, P., Boerlage, M., Steinbuch, M.: Trajectory Planning and Feedforward Design for Electromechanical Motion Systems. Control Engineering Practice 13(2), 145–157 (2005)

[16] Longman, R.W.: Iterative Learning Control and Repetitive Control for Engineering Practice. International Journal of Control 73(10), 930–954 (2000)

[17] Van der Meulen, S.H., Tousain, R.L., Bosgra, O.H.: Fixed Structure Feedforward Controller Design Exploiting Iterative Trials: Application to a Wafer Stage and a Desktop Printer. Journal of Dynamic Systems, Measurement, and Control 130(051006) (2008)

[18] Moore, K.L.: Iterative learning control: An expository overview. In: B.N. Datta (ed.) Applied and Computational Control, Signals, and Circuits, vol. 1, chapter 4, pp. 151–214. Birkhäuser, Boston (1999)

[19] Nash, S.G., Sofer, A.: Linear and Nonlinear Programming. McGraw-Hill Series in Industrial Engineering and Management Science. McGraw-Hill, London (1996)

[20] Phan, M., Longman, R.W.: A Mathematical Theory of Learning Control for Linear Discrete Multivariable Systems. In: Proceedings of the AIAA/AAS Astrodynamics Conference, pp. 740–746. Minneapolis, Minnesota (1988)

[21] Steinbuch, M., Norg, M.L.: Advanced Motion Control: An Industrial Perspective. European Journal of Control 4(4), 278–293 (1998)

[22] Stix, G.: Trends in Semiconductor Manufacturing: Toward "Point One". Scientific American 272(2), 72–77 (1995)

[23] Tomizuka, M.: Zero Phase Error Tracking Algorithm for Digital Control. Journal of Dynamic Systems, Measurement, and Control 109(1), 65–68 (1987)

[24] Torfs, D.E., Vuerinckx, R., Swevers, J., Schoukens, J.: Comparison of Two Feedforward Design Methods Aiming at Accurate Trajectory Tracking of the End Point of a Flexible Robot Arm. IEEE Transactions on Control Systems Technology 6(1), 2–14 (1998)

[25] Tousain, R., Van der Meché, E., Bosgra, O.: Design Strategy for Iterative Learning Control Based on Optimal Control. In: Proceedings of the 40th IEEE Conference on Decision and Control, vol. 5, pp. 4463–4468. Orlando, Florida (2001)

[26] Tousain, R., Van Casteren, D.: Iterative Learning Control in a Mass Product: Light on Demand in DLP projection systems. In: Proceedings of the 2007 American Control Conference, pp. 5478–5483. New York City, New York (2007)

[27] Tsao, T.C., Tomizuka, M.: Robust Adaptive and Repetitive Digital Tracking Control and Application to a Hydraulic Servo for Noncircular Machining. Journal of Dynamic Systems, Measurement, and Control 116(1), 24–32 (1994)

[28] Van de Wal, M., Van Baars, G., Sperling, F., Bosgra, O.: Multivariable $\mathcal{H}_\infty/\mu$ Feedback
     Control Design for High-Precision Wafer Stage Motion. Control Engineering Practice **10**(7),
     739–755 (2002)
[29] Zhao, S., Tan, K.K.: Adaptive Feedforward Compensation of Force Ripples in Linear Motors.
     Control Engineering Practice **13**(9), 1081–1092 (2005)

Incremental Identification of Hybrid Models of Dynamic Process Systems

Olaf Kahrs, Marc Brendel, Claas Michalik and Wolfgang Marquardt

Abstract This contribution presents the so called incremental approach to the general modeling task and shows various fields of application as well as conceptual extensions of the method. The incremental model identification procedure has been developed within a collaborative interdisciplinary research center (CRC) at RWTH Aachen. First, the so called MEXA process, which is at the core of the research at the CRC is presented. Next, the incremental model identification approach (which is one crucial step within the MEXA process) is contrasted with the classical simultaneous approach. The application of the incremental approach is then shown for the special case of hybrid reaction kinetic models. In a next step, the basic idea of the incremental approach - the decomposition of the problem into simpler subproblems - is generalized to also account for (mechanistic and hybrid) algebraic and dynamic models (from arbitrary fields, e.g., not necessarily reaction kinetics). Finally, open questions within the incremental framework are discussed and the future research focus is given.

Olaf Kahrs
Aachener Verfahrenstechnik - Process Systems Engineering, RWTH Aachen University, Germany.
Current affiliation: BASF SE, Ludwigshafen, Germany. e-mail: olaf.kahrs@basf.com

Marc Brendel
Aachener Verfahrenstechnik - Process Systems Engineering, RWTH Aachen University, Germany.
Current affiliation: Evonik Degussa GmbH, Hanau, Germany. e-mail: marc.brendel@evonik.com

Claas Michalik
Aachener Verfahrenstechnik - Process Systems Engineering, RWTH Aachen University, Germany.
e-mail: claas.michalik@avt.rwth-aachen.de

Wolfgang Marquardt
Aachener Verfahrenstechnik - Process Systems Engineering, RWTH Aachen University, Germany.
e-mail: wolfgang.marquardt@avt.rwth-aachen.de

P.M.J. Van den Hof et al. (eds.), *Model-Based Control: Bridging Rigorous Theory and Advanced Technology*, DOI: 10.1007/978-1-4419-0895-7_11,
© Springer Science + Business Media, LLC 2009

1 Introduction

Within the last decades process models have been increasingly utilized for various applications in industrial practice. Typical applications include safety analysis of process plants, the development of soft sensors as well as real-time process optimization and model-predictive control. The continuously increasing computational power facilitates the handling of more and more complex process models, describing coupled systems of various production and or separation units on a very detailed level of accuracy covering not only lumped process systems in thermodynamical equilibrium but also distributed parameter systems governed by kinetic reaction and transport processes [19]. However, due to limited process knowledge, detailed mechanistic modeling of kinetic phenomena is often not feasible for all submodels describing the models of the process units of the plant. Therefore, the concept of hybrid or gray-box modeling – combining mechanistically motivated and purely data-driven model parts – becomes increasingly important [20, 1].

Though more demanding - especially in terms of the available measurement data - the identification and discrimination process for hybrid modeling roughly follows the same steps as the identification of a mechanistic model. A systematic approach to model identification and discrimination has been developed within the Collaborative Research Center "Model Based Experimental Analysis" (or MEXA for short) which has been funded as CRC 540 by the DFG, the German Science Foundation, from 1999 to 2009 (see: http://www.sfb540.rwth-aachen.de). The MEXA work process [16] aims at a tight integration of the steps carried out during model identification including

- optimal experimental design to determine those experimental conditions yielding measurements with maximum information content,
- model building and refinement,
- the solution of inverse problems to fit a model to the data, and
- methods for selecting the most suitable model structure and for model falsification.

While each of these topics constitutes an active field of research itself, a structured and efficient method for the overall model identification process is still lacking. Therefore the MEXA work process aims at filling this gap. The most important steps within the MEXA work process are the solution of the inverse problems to fit a set of alternative model candidates for the individual submodels of the process plant model to experimental data and to decide which of these submodels best explain the experimental data (model identification and discrimination). A novel, efficient decomposition strategy has been developed for the solution of these demanding inverse problems as part of the research of CRC 540, originally aiming at mechanistic modeling. Before we show how this so-called *incremental identification method* [3] can be extended to allow the solution of the inverse problems associated with the identification and discrimination of hybrid models as discussed in more detail in Sections 2 to 5, we present the general concept and contrast it with the classical *simultaneous identification method* in Section 3 after a short introduction to hybrid models in Section 2.

2 Hybrid models

If a model is desired for a process that involves complex kinetic phenomena, for example biochemical reactions or polymerization reactions, it becomes increasingly difficult to specify a suitable set of candidate models for model identification and discrimination. Furthermore, when the set of model candidates does not include the correct kinetic model structure, a model mismatch will result.

For these processes the hybrid modeling approach is advantageous. In a hybrid model those model parts are described by empirical models, such as neural nets or polynomials, for which only limited physico-chemical understanding is available. Since these empirical models can be considered as universal approximators (if carefully chosen), this means they can approximate nonlinear functions arbitrarily well, a model mismatch can be avoided in principle.

However, the increased flexibility of the empirical model parts in a hybrid model requires tailored model identification and validation approaches. The identification approach has to determine a suitable set of empirical model variables, the empirical model structure (such as the number of basis functions), and has to estimate the parameters in the respective empirical models [5, 11]. The validity range of the empirical models should be monitored during model application to prevent incorrect hybrid model predictions [12].

3 Model identification strategies

To illustrate the classical, simultaneous model identification process, consider for example the modeling of a two-phase distributed enzyme catalyzed reaction system. Let us therefore assume that the enzyme is immobilized in hydrogel beads, which are located in a well-mixed liquid solvent that serves as a reservoir for the substrates and reaction products. Assuming an isothermal reaction, a model of the reactor would require kinetic models for (i) the mass transfer between the two phases, (ii) the diffusion within the hydrogel bead and for (iii) the reaction rates in the hydrogel. For each kinetic phenomenon several alternative model structures based on different assumptions and theories may exist. The aggregation of such sub-models will inevitably lead to a quickly (i.e. combinatorially) growing number of aggregated candidate models for the two-phase bioreactor. Usually, a first fit of all the unknown parameters of the originally proposed submodel structures is not satisfactory in the light of the measurement data and some model improvement strategy has to be selected. Typically, such model improvement is done by a process expert based on his professional intuition. For example, the process expert may suspect some submodel structure to be responsible for the bad fit. Consequently, the submodel under consideration is replaced with another which seems to be more adequate and the parameter estimation process is repeated. Usually, there is more than one submodel to be replaced and a systematic procedure of replacing or modifying the submodels and their combination is lacking. Rather, an ad hoc improvement approach is implemented by the practitioner, where the number of aggregated model structures is quickly growing due to the nested submodels and the resulting combinatorial nature of the model selection problem. Obviously, a more structured modeling strategy is desirable to select and validate a model structure on scientific grounds reflecting experimental evidence.

3.1 Incremental model development and refinement

Before continuing the discussion on model identification, we want to recall that the development of the model structure itself, e.g. the collection of the model equations, can be carried out in a systematic manner [15]. In a first modeling step,

- the balance envelopes are chosen and their interactions are determined,
- the intended spatio-temporal resolution within the balance envelopes and hence of the model is decided,
- and extensive quantities are selected for balancing.

In case of the already mentioned bioreactor with enzymes immobilized in hydrogel beads, well-mixed fermentation broth and the collection of spatially distributed hydrogel beads, interacting through a common interface, are chosen as suitable balance envelopes. Mass is selected to be balanced. Subsequently, the balance equations are formulated and gradually refined as illustrated in Fig. 1. On level B the balance is formulated as a sum of generalized fluxes, i.e. the inter- and intraphase transport as well as the source/sink terms. On the next decision level BF, constitutive equations are specified for each flux term in the balances, i.e. the models for the interfacial mass fluxes, the diffusive fluxes within the hydrogel bead as well as the bioreaction rates in the bioreactor example considered. Often, these kinetic submodels depend on both, thermodynamic state functions (i.e. density, heat capacity etc.) and rate coefficients (i.e. reaction rate, diffusion or mass transfer coefficients etc.), which themselves depend on the thermodynamic state. Consequently, the structure of the expression relating rate coefficients and states is located on yet another level BFR. This *incremental model building* serves as a structured road-map for modeling purposes.

The submodel structures chosen on the decision levels BF and BFR do not necessarily have to be based on first principles. In particular, any mathematical description can be selected to express the correlation of a flux or a kinetic coefficient and the intensive thermodynamical state variables. If at least one of the submodels is chosen to comprise an in general multivariate function, a hybrid (or grey-box) model [20, 18, 11] naturally arises, since first principles models and empirical models are

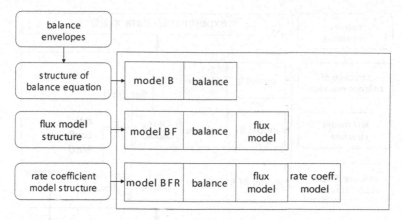

Fig. 1 Incremental refinement during the development of model equations

combined on different decision levels. In particular, the balance equations on level B are always based on first principles while any of the flux and rate coefficient models on levels BF and BFR can be either purely data-driven or rigorous.

3.2 Incremental model identification

The structured and systematic model building process described in the previous sub-section is also a good starting point for a structured process for the identification of a suitable model structure consisting of aggregated submodels. The resulting work process is entitled *incremental model identification* and depicted in Fig. 2. For sim-plicity, we assume to have access to experimental data for the states $x(z,t)$ which are considered to depend on time t and a single spatial coordinate z at sufficient spatial and temporal resolution. Hence - at least in principle - the unknown flux $J(z,t)$ in the balance equation on level B can be estimated based on these experimental data. Note that the flux can be estimated without the need for specifying a constitutive equation. On the next level, BF a rate estimate $k(z,t)$ can be obtained based on the flux estimates together with the measurements. Often, the flux model can directly be solved for the rate coefficient function $k(z,t)$. Finally, on level BFR which is as-sumed to only depend on the measured states and constant parameters, a model for the rate coefficients is identified. These parameters of the rate model can be com-puted from the estimated rate coefficients $k(z,t)$ and the measured states $x(z,t)$ by the solution of an algebraic (potentially nonlinear) parameter estimation problem.

  The incremental identification approach generalizes the so-called *differential method* of reaction kinetics [9], where the reaction fluxes are measured or inferred from inferential measurements and correlated with measured concentrations either by regression [22, 7, 14] or by mechanistically motivated [23] models.

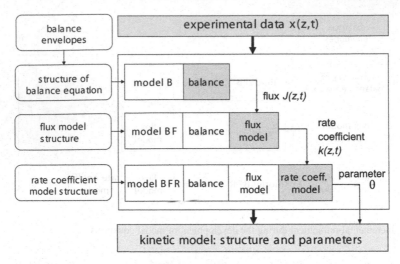

Fig. 2 Incremental refinement during the identification of a model structure

3.3 *An assessment of incremental identification*

The incremental approach, as presented above, has the potential to (at least partially) overcome the disadvantages of the simultaneous approach. Due to the decomposition of one large problem into a sequence of smaller, less complex ones, the overall complexity is reduced drastically. In addition, the combinatorial explosion of model alternatives, due to the necessary combination of various submodels, each associated with a number of modeling alternatives, is avoided. This is achieved because the submodels can be judged independently of one another, such that inadequate model parts can also be identified directly. Hence, the ad hoc and intuition-driven model refinement, which is commonly used in academia as well as in industry, is replaced by a well-organized strategy.

The incremental identification strategy also offers computational advantages. The computationally very demanding solution of many difficult least-squares problems with differential-algebraic or even partial differential-algebraic constraints and a possibly large number of experimental data points resulting from high resolution measurements is completely avoided. Rather, a sequence of

- typically a linear inverse problem involving differential equations in the first step and
- (non)linear regression problems with typically algebraic constraints in the remaining steps

has to be solved. This decomposition facilitates initialization and convergence of the estimation algorithms and reduces the computational effort drastically. Intractable estimation problems (such as those involving distributed parameter systems with many point measurements) may thus become computationally tractable.

4 Incremental identification of a hybrid semi-batch reactor

In this section, we illustrate the hybrid modeling approach following the incremental identification strategy by means of a case study involving a semi-batch reactor with a homogeneous liquid reactor content [5]. We assume different amounts of prior knowledge, leading to three identification scenarios. For all scenarios, we assume that no mechanistically motivated model structure for the reaction rates are present. Hence, these rates are modeled in a data-driven manner using feed-forward neural networks.

4.1 Reactor model

Consider a homogeneous, open chemical reaction system where S species involved in R reactions are present. The number of moles of species i, n_i [mol], is given by

$$\frac{dn_i}{dt} = f_i^{in} - f_i^{out} + f_i^{r}, \quad i = 1, .., S, \tag{1}$$

where f_i^{in} and f_i^{out} [mol/min] are the molar flow rates of species i into and out of the reactor and f_i^{r} [mol/min] is the reaction flux of species i, i.e. the net molar flow rate of species i produced or consumed by the various chemical reactions.

The reaction flux of species i can be expressed in terms of the individual reaction rates to result in

$$f_i^{r} = V \sum_{j}^{R} v_{ij} r_j, \quad i = 1, .., S \tag{2}$$

where v_{ij} refers to the stoichiometric coefficient for species i in the j-th reaction, r_j [mol/l min] to the rate of the j-th reaction, and V [l] to the volume.

For a constant-density reaction fluid, a volumetric feed of rate F [l/min] and concentration c_i^{in} [mol/l], vanishing outflow, the mole balance equation (1) expressed in terms of the molar concentration $c_i = n_i/V$ [mol/l], and the total mass balance of the semi-batch reactor can be written as

$$\frac{dc_i}{dt} = \frac{F}{V}(c_i^{in} - c_i) + \frac{f_i^{r}}{V} \tag{3}$$

$$\frac{dV}{dt} = F, \tag{4}$$

if the volume does not change due to chemical reactions or non-ideal thermodynamic liquid properties.

192   Olaf Kahrs, Marc Brendel, Claas Michalik and Wolfgang Marquardt

4.2 Incremental identification approach for the hybrid semi-batch reactor example

The incremental identification approach, as described in general terms in Section 3.2, includes the following steps for the reactor model considered here:

1) The fluxes $\hat{f}_i^r(t), i=1,..,S$ are estimated using mole balances (model 1 on level B) and noisy quantities $\bar{n}_i = \bar{c}_i V$.
2) With additional information on stoichiometry (model 2 on level BF), the reaction rates $\hat{r}_j(t)$, j=1,..,R are then calculated using (2).
3) Furthermore, if the rate laws (e.g. $r = kc_A c_B$) are known (model 3 on level BFR), (time-variant) rate constants $\hat{k}_j(t)$ are calculated from the reaction rate $\hat{r}_j(t)$ and concentrations $\hat{c}_i(t)$.
4) If, in addition, a temperature dependency of k should occur (model 4), such as the Arrhenius law ($k = k_0 e^{\frac{-E}{RT}}$), the rate constant parameters ($\hat{k}_{0j}, \hat{E}_j$) can then be estimated from $\hat{k}_j(t)$ and $\hat{T}(t)$ on an additional refinement level following BFR.

For the last case - an unknown temperature dependency - the outputs of model 3 can be taken as inputs to a data-driven approach for describing $k = k(T)$. For an unknown rate law (model 2 known), $\hat{c}_i$, T and $\hat{r}_j$ serve as inputs to the data-driven models $r_j = r_j(c_i)$ and $r_j = r_j(c_i, T)$ for the isothermal and non-isothermal cases, respectively. If the reaction stoichiometry is unknown, target factor analysis [4] can help to verify hypotheses on the stoichiometry based on the estimated fluxes.

4.3 Simulated isothermal reaction system

The incremental approach for identifying reaction kinetics is illustrated on the aceto-acetylation of pyrrole with diketene [21]:

$$P+D \xrightarrow{K} PAA \qquad (5a)$$

$$D+D \xrightarrow{K} DHA \qquad (5b)$$

$$D \rightarrow oligomers \qquad (5c)$$

$$PAA+D \xrightarrow{K} F \qquad (5d)$$

In addition to the desired reaction of diketene (D) and pyrrole (P) to 2-acetoacetyl pyrrole (PAA) (5a), there are several undesired side reactions (5b)-(5d). The reactions take place isothermally in a laboratory-scale semi-batch reactor with an initial volume of 1 liter. A diluted solution of diketene is added continuously. The symbol $\xrightarrow{K}$ indicates that a reaction is catalyzed by pyridine (K), the concentration of which continuously decreases during the run due to addition of the diluted diketene feed. The dilution of catalyst is modeled by normalizing the corresponding rate constants with respect to the volume. Hence, the effective reaction rates are described by the constitutive equations

$$r_j(t) = \frac{V_0}{V(t)} r_j^\star(t), \quad j = \{a,b,d\}, \tag{6}$$

$$r_c(t) = r_c^\star(t), \tag{7}$$

with the reaction rate laws

$$r_a^\star(t) = k_a c_P(t) c_D(t), \tag{8a}$$

$$r_b^\star(t) = k_b c_D^2(t), \tag{8b}$$

$$r_c^\star(t) = k_c c_D(t), \tag{8c}$$

$$r_d^\star(t) = k_d c_{PAA}(t) c_D(t), \tag{8d}$$

where V_0 represents the initial volume and k_a, k_b, k_c and k_d the rate constants.

The mole balances for the species D, P, PAA and DHA read

$$\frac{dc_D(t)}{dt} = \frac{F(t)}{V(t)}[c_D^{in} - c_D(t)] + \frac{f_D^r(t)}{V(t)}, \tag{9a}$$

$$\frac{dc_P(t)}{dt} = -\frac{F(t)}{V(t)} c_P(t) + \frac{f_P^r(t)}{V(t)}, \tag{9b}$$

$$\frac{dc_{PAA}(t)}{dt} = -\frac{F(t)}{V(t)} c_{PAA}(t) + \frac{f_{PAA}^r(t)}{V(t)}, \tag{9c}$$

$$\frac{dc_{DHA}(t)}{dt} = -\frac{F(t)}{V(t)} c_{DHA}(t) + \frac{f_{DHA}^r(t)}{V(t)}, \tag{9d}$$

with initial conditions $c_D(0) = c_{D0}$, $c_P(0) = c_{P0}$, $c_{PAA}(0) = c_{PAA0}$ and $c_{DHA}(0) = c_{DHA0}$. The reaction fluxes f_D^r, f_P^r, f_{PAA}^r and f_{DHA}^r can be related to the reaction rates using the stoichiometry as follows:

$$f_D^r = (-r_a - 2r_b - r_c - r_d)V, \tag{10a}$$

$$f_P^r = -r_a V, \tag{10b}$$

$$f_{PAA}^r = (r_a - r_d)V, \tag{10c}$$

$$f_{DHA}^r = r_b V. \tag{10d}$$

4.4 Experimental data

To allow an easy interpretation of the results, concentration trajectories are generated using the model described above and the rate constants given in Table 1.

The measured concentrations are taken at a sampling frequency $f_s = 60$ min^{-1} and are corrupted with normally distributed white noise of standard deviation $\sigma_c = 0.01$ mol/l. This corresponds to an absolute error of approximately 2%. The

Table 1 Values of rate constants

	k_a $[\frac{l}{mol\,min}]$	k_b $[\frac{l}{mol\,min}]$	k_c $[\frac{1}{min}]$	k_d $[\frac{l}{mol\,min}]$
value	0.053	0.128	0.028	0.001

batch time is $t_f = 60$ min. Only the species D, P, PAA and DHA are assumed to be measured.

To obtain reliable approximations of the reaction rates, experiments are designed such that the experimental data cover a large domain of the concentration state space. Six independent variables can be considered: the four initial conditions c_{D0}, c_{P0}, c_{PAA0} and c_{DHA0}, the constant feed rate F, and feed concentration c_D^{in}. The possible ranges for these experimental degrees of freedom are given in Table 2. Since the transient behavior is most sensitive with respect to c_{D0}, c_{P0}, F and c_D^{in}, a 2^{6-2} factorial design consisting of 16 experiments is selected. A more elaborate procedure to determine the initial experiments aiming at a balanced covering of the concentration space has been reported elsewhere [6].

Table 2 Range of independent variables

	c_{D0} $[\frac{mol}{l}]$	c_{P0} $[\frac{mol}{l}]$	c_{PAA0} $[\frac{mol}{l}]$	c_{DHA0} $[\frac{mol}{l}]$	F $[\frac{l}{min}]$	c_D^{in} $[\frac{mol}{l}]$
min	0.07	0.40	0.10	0.02	0.5e-3	4.0
max	0.14	0.80	0.20	0.04	1.5e-3	6.0

4.5 Various modeling scenarios

In the following, three different modeling scenarios are presented, each differing in the amount of prior knowledge regarding the reactions. The fluxes, reaction rates and reaction kinetics are identified from noise-corrupted, simulated concentration data.

In the first scenario, we are aware of the existence and stoichiometry of reactions (5a)-(5d) and also know the effect of the catalyst on reactions (5a), (5b) and (5d). Due to the knowledge on the structure of the reaction mechanism, the formal reaction rates $r_j^\star$ can be calculated in this scenario. Hence, the concentrations and the reaction rates can be correlated as $r_a^\star = r_a^\star(c_P, c_D)$, $r_b^\star = r_b^\star(c_D)$, $r_c^\star = r_c^\star(c_D)$ and $r_d^\star = r_d^\star(c_{PAA}, c_D)$, using a feed-forward neural net with Bayesian regularization as training algorithm and 3 nodes in the hidden layer.

In the second scenario, the knowledge regarding the effect of the catalyst on the kinetics is suppressed, otherwise, the procedure is identical to that of the first scenario.

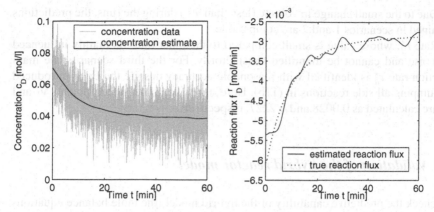

Fig. 3 True and estimated reaction flux f_D^r (right), measured and estimated concentration c_D (left)

In the third scenario, besides the known desired reaction (5a), there is only evidence that D and P are involved in other reactions, including the formation of dehydroacetic acid DHA. Hence, the lumped stoichiometric model

$$P+D \rightarrow PAA \tag{11a}$$

$$D + \nu_1 PAA \rightarrow \nu_2 DHA + G \tag{11b}$$

with the unknown stoichiometric coefficients ν_1 and ν_2 is postulated. From the estimated reaction fluxes, the reaction rates $r_a^\star(t)$ for Reaction (11a) and $r_{lump}^\star(t)$ for Reaction (11b) as well as the stoichiometric coefficients ν_1 and ν_2 can be determined as solution of the reconciliation problem

$$f_D^r(t) = [-r_a^\star(t) - r_{lump}^\star(t)]V(t) \tag{12a}$$

$$f_P^r(t) = -r_a^\star(t)V(t) \tag{12b}$$

$$f_{PAA}^r(t) = [r_a^\star(t) - \nu_1 r_{lump}^\star(t)]V(t) \tag{12c}$$

$$f_{DHA}^r(t) = \nu_2 r_{lump}^\star(t)V(t). \tag{12d}$$

The rates $r_a^\star(t)$ and $r_{lump}^\star(t)$ can subsequently be correlated with the concentrations, as discussed in Scenario 1.

Exemplarily, the true and estimated reaction fluxes for species D are shown in Figure 3 (right). Integration of (9a) yields an estimate of the concentration c_D, as shown in Figure 3 (left).

To demonstrate the quality of the resulting models in Table 3 we compare the mean and maximum values of the neural net predictions in the validity range[1].

---

[1] The smallest n-dimensional (e.g. $n = 2$ for the bivariate case) box containing all concentration combinations taken for training

Due to the small change in volume (less than 9%) during the runs, the predictions obtained in scenarios 1 and 2 are comparable.

Rate r_d^*, whose value is small compared to the other rates, is mainly influenced by noise and cannot be identified satisfactorily. For the third scenario, the main reaction rate r_a^* is identified with reasonable accuracy despite the error introduced by lumping all side reactions in (11b). Here, the stoichiometric coefficients v_1 and v_2 are calculated as 0.0028 and 0.2227, respectively.

4.6 Validation of the hybrid reactor model

To check the predictive capability of the hybrid model (the mole balance equations and the neural-net-based kinetic laws), concentration trajectories are simulated using (8a)-(8d) and compared to the hybrid model predictions for ten runs with experimental conditions chosen randomly within the ranges given in Table 2. The mean and maximum values of the prediction errors are listed in Table 4.

It can be seen that all hybrid models perform excellent on this test set. Even in case of a poorly estimated reaction rate r_d^*, the overall prediction is accurate, which demonstrates the low sensitivity of this rate and hence also explains why it cannot be estimated accurately.

Table 3 Reaction rate prediction errors

		r_a^*	r_b^*	r_c^*	r_d^*
Scenario 1	Mean error	2.95	6.15	5.16	185
	Max. error	11.24	26.42	20.92	3245
Scenario 2	Mean error	4.48	7.33	4.88	117
	Max. error	15.69	25.84	18.58	466
Scenario 3	Mean error	4.64	-	-	-
	Max. error	62.99	-	-	-

Table 4 Hybrid model prediction errors

	[%]	c_D	c_P	c_{PAA}	c_{DHA}
Scenario 1	Mean error	0.81	0.11	0.40	1.10
	Max. error	2.80	0.63	1.48	5.58
Scenario 2	Mean error	0.79	0.24	0.61	1.10
	Max. error	2.30	0.43	1.19	5.04
Scenario 3	Mean error	0.60	0.20	0.53	4.13
	Max. error	1.91	0.42	1.28	15.58

5 Incremental identification of generally structured hybrid models

When the process under investigation contains several interconnected process units (reactors, mixers, distillation columns, etc.) the number of model equations that have to be considered during model identification becomes large. Especially for these large models, an incremental identification approach can significantly reduce the identification effort. However, the specific structure of the model exploited in the semi-batch reactor case study is lost and the more general structure of the overall process model requires a generalization of the incremental identification approach. Such an extension has been presented in detail in [11] for stationary problems involving algebraic models only. We generalize the idea reported there towards incremental identification of differential-algebraic models in this section.

The underlying strategy in the incremental approach presented in Section 3.1 is to estimate unknown quantities from a minimal subset of equations in a stepwise manner. The subset of equations usually contains measured or previously estimated quantities. This identification strategy is generalized by the algorithm of [11] which decomposes an algebraic hybrid model into subsets of model equations that can be solved sequentially for the unknown quantities. Each subproblem is analyzed and solved by one of the following methods: data reconciliation, nonlinear equation solving, parameter estimation, or empirical model identification. Finally, a correction step is carried out by the simultaneous identification approach to obtain statistically optimal parameter estimates.

In the following, we restrict ourselves to differential-algebraic models that can be rephrased to algebraic models by suitable algorithms [17, 13] if estimates of the derivatives of the states occurring in the model are available. It is important to mention that the estimation of derivatives based on noisy data is an ill-posed problem and requires appropriate regularization strategies. Different regularization methods are discussed elsewhere [10].

In the simultaneous identification approach, the unknown parameters $\mathbf{p} \in \mathbb{R}^{m_p}$ of the algebraic hybrid model

$$\mathbf{f}(\mathbf{x}, \mathbf{y}, \mathbf{p}) = \mathbf{0}, \tag{13}$$

with $\mathbf{f} : (\mathbb{R}^{m_x} \times \mathbb{R}^{m_y} \times \mathbb{R}^{m_p}) \rightarrow \mathbb{R}^{m_f}$, are estimated from the measured process variables $\tilde{\mathbf{y}}_i \in \mathbb{R}^{m_y}$, where $i = 1, \ldots, n$ denotes the i-th data point in the measurement data set. Note that the measured data set does not only include the actual measurements but also the estimates of the time derivatives of the measured quantities. For simplicity, all these quantities are referred to as measurements subsequently. For the measured variables we assume $\tilde{\mathbf{y}}_i = \mathbf{y}_i^* + \varepsilon_i$, where $\mathbf{y}_i^*$ denotes the noise-free, but unknown value of the measured variables. The measurement error ε_i is assumed to be normally distributed with zero mean and covariance matrix V. Furthermore, the hybrid model also contains the non-measured process variables $\mathbf{x} \in \mathbb{R}^{m_x}$ which may also constitute of time-derivatives if they show up in the original dynamic model but are not accessible from differentiating the measured process variables.

The parameters $\mathbf{p}$ are estimated by solving the nonlinearly constrained weighted least-squares problem

$$\min_{\mathbf{x}_i, \mathbf{y}_i, \mathbf{p}} \quad \frac{1}{2} \sum_{i=1}^{n} (\mathbf{y}_i - \tilde{\mathbf{y}}_i)^T W (\mathbf{y}_i - \tilde{\mathbf{y}}_i) + \lambda R(\mathbf{p}) \tag{14}$$

$$s.t. \quad \mathbf{f}(\mathbf{x}_i, \mathbf{y}_i, \mathbf{p}) = \mathbf{0}$$
$$\mathbf{g}(\mathbf{x}_i, \mathbf{y}_i, \mathbf{p}) \leq \mathbf{0}$$
$$\mathbf{x}_l \leq \mathbf{x}_i \leq \mathbf{x}_u, \quad \mathbf{y}_l \leq \mathbf{y}_i \leq \mathbf{y}_u, \quad \mathbf{p}_l \leq \mathbf{p} \leq \mathbf{p}_u$$
$$\forall \quad i = 1, \dots, n,$$

with the weighting matrix W, and lower and upper bounds on the process variables and parameters. If the weighting matrix is chosen to be the inverse of the covariance matrix V and if the hybrid model does not contain structural errors, the weighted least-squares problem can be derived within a maximum likelihood framework and yields statistically sound parameter estimates [2].

In the incremental identification approach, the parameter estimation problem in Eq. (14) is decomposed into a set of simpler problems, whose solution is an approximation of the solution of Eq. (14). If estimates of the measured variables $\hat{\mathbf{y}}_i$ would be available without minimizing Eq. (14) and $nm_f \geq nm_x + m_p$, then the non-measured variables $\mathbf{x}_i$ and the parameters $\mathbf{p}$ could be estimated by solving the constraints $\mathbf{f}(\mathbf{x}_i, \hat{\mathbf{y}}_i, \mathbf{p}) = \mathbf{0}, \forall \quad i = 1, \dots, n$. Estimated values are denoted by $\hat{\ }$. Since $\hat{\mathbf{y}}_i$ is not known prior to solving Eq. (14), we assume in the incremental approach $\hat{\mathbf{y}}_i \approx \tilde{\mathbf{y}}_i$. Thus, measured values $\tilde{\mathbf{y}}_i$ are used to estimate the non-measured variables and parameters by solving $\mathbf{f}(\mathbf{x}_i, \tilde{\mathbf{y}}_i, \mathbf{p}) = \mathbf{0}, \forall \quad i = 1, \dots, n$. Instead of solving this usually overdetermined system of nonlinear equations ($nm_f > nm_x + m_p$), we exploit its structure and decompose it into a sequence of simpler problems.

The hybrid model decomposition algorithm is based on the incidence matrix $A \in \mathbb{R}^{nm_f} \times \mathbb{R}^{nm_x + nm_y + m_p}$ of the equality constraints $\mathbf{f}(\bullet)$ of Eq. (14), whose entries A_{ij} are 1 if the i-th constraint contains the j-th unknown optimization variable and zero otherwise, with $i = 1, \dots, nm_f$ and $j = 1, \dots, nm_x + nm_y + m_p$. This incidence matrix is used by the Dulmage-Mendelsohn (DM) algorithm presented in (**author?**) [8] to decompose the system of nonlinear equations into an under-determined, a determined, and an over-determined subsystem.

The DM algorithm is utilized for decomposing the equality constraints in Eq. (14), as described in the following:

1) First, the equality constraint incidence matrix A of Eq. (14) with dimension $nm_f \times (nm_x + nm_y + m_p)$ is reduced to a $nm_f \times (nm_x + m_p)$ dimensional incidence matrix by substituting known measurement data $\tilde{\mathbf{y}}_i$ for the variables $\mathbf{y}_i$. Thus, the measured variables $\mathbf{y}_i$ are specified as known variables and are dropped from the set of optimization variables.
2) Since the parameters have to be estimated from the measured variable values of all n data points, the system of equality constraints $\mathbf{f}(\cdot)$ can further be decomposed into two parts: a part $\mathbf{f}^k(\cdot)$ that is independent of the unknown parameters $\mathbf{p}$ and therefore called the *known* part, and an *unknown* part $\mathbf{f}^u(\cdot)$ that

contains all equations depending on the unknown parameters. After the known part $\mathbf{f}^k(\mathbf{x}_i^k, \tilde{\mathbf{y}}_i^k) = \mathbf{0}$ is solved for all $i = 1, \ldots, n$ data points separately, the estimated values $\hat{\mathbf{x}}_i^k$ and $\hat{\mathbf{y}}_i^k$ are inserted into the unknown part $\mathbf{f}^u(\hat{\mathbf{x}}_i^k, \hat{\mathbf{y}}_i^k, \mathbf{x}_i^u, \tilde{\mathbf{y}}_i^u, \mathbf{p}) = \mathbf{0}$ to estimate the unknown parameters $\mathbf{p}$ and variables $\mathbf{x}_i^u$ and $\mathbf{y}_i^u$.

3) The known part can be further decomposed by the DM algorithm into an under-determined, determined, and over-determined part, denoted by superscripts ^{ku}, ^{kd}, and ^{ko}, respectively. Since the sets of process variables $\mathbf{x}_i^k$ and $\mathbf{y}_i^k$ are disjunct for all $i = 1, \ldots, n$ data points, the DM algorithm can be applied for each data point independently. Furthermore, if the respective parts of the incidence matrix are identical (i.e., if there are no missing values for some data points), then the decomposition is identical for all data points and has to be calculated only once. When *solving a system of nonlinear equations*, the under-determined and the over-determined sets of equations have to be empty. If this is not the case, the model has to be analyzed for redundant or missing equations, and for wrong specification of the set of unknown variables. In *parameter estimation*, the over-determined part indicates redundant measured variables (when no modeling errors are present). Therefore, the over-determined part is interpreted as a data reconciliation problem and is solved accordingly. The determined part contains as many equations as unknown variables and has full structural rank, such that the system of equations is solved for the unknown variables. In the under-determined part, the number of equations is lower than the number of unknown variables and has to be solved together with the unknown part containing the parameters.

4) Finally, the combined unknown and under-determined part as well as the over-determined part are further decomposed into independent parts with disjunct variable and parameter sets.

This procedure decomposes the hybrid model into four major parts, namely

$$\mathbf{f}^{ko}(\mathbf{x}_i^{ko}, \tilde{\mathbf{y}}_i^{ko}) = \mathbf{0}, \tag{15}$$

$$\mathbf{f}^{kd}(\hat{\mathbf{x}}_i^{ko}, \hat{\mathbf{y}}_i^{ko}, \mathbf{x}_i^{kd}, \tilde{\mathbf{y}}_i^{kd}) = \mathbf{0}, \tag{16}$$

$$\mathbf{f}^{ku}(\hat{\mathbf{x}}_i^{ko}, \hat{\mathbf{y}}_i^{ko}, \hat{\mathbf{x}}_i^{kd}, \tilde{\mathbf{y}}_i^{kd}, \mathbf{x}_i^{ku}, \tilde{\mathbf{y}}_i^{ku}) = \mathbf{0}, \tag{17}$$

$$\mathbf{f}^u(\hat{\mathbf{x}}_i^{ko}, \hat{\mathbf{y}}_i^{ko}, \hat{\mathbf{x}}_i^{kd}, \tilde{\mathbf{y}}_i^{kd}, \mathbf{x}_i^{ku}, \tilde{\mathbf{y}}_i^{ku}, \mathbf{x}_i^u, \tilde{\mathbf{y}}_i^u, \mathbf{p}) = \mathbf{0}. \tag{18}$$

Each of these sets of nonlinear equations may show a particular substructure which could be exploited by a numerical solution method.

The four sets of equations are solved subsequently, starting with the over-determined part. The unknown variables $\mathbf{x}_i^{ko}, i = 1, \ldots, n$ are estimated by performing a data reconciliation with the over-determined rigorous part $\mathbf{f}^{ko}(\cdot) = \mathbf{0}$ as constraints for each data point separately:

$$\min_{\mathbf{x}_i^{ko},\mathbf{y}_i^{ko}} \frac{1}{2}(\mathbf{y}_i^{ko} - \tilde{\mathbf{y}}_i^{ko})^T W^{ko}(\mathbf{y}_i^{ko} - \tilde{\mathbf{y}}_i^{ko}) \tag{19}$$

$$s.t. \quad \mathbf{f}^{ko}(\mathbf{x}_i^{ko},\mathbf{y}_i^{ko}) = 0$$

$$\mathbf{x}_l^{ko} \le \mathbf{x}_i^{ko} \le \mathbf{x}_u^{ko}, \quad \mathbf{y}_l^{ko} \le \mathbf{y}_i^{ko} \le \mathbf{y}_u^{ko}.$$

The estimated variables $\hat{\mathbf{x}}_i^{ko}$ and $\hat{\mathbf{y}}_i^{ko}$ are inserted into the determined rigorous part $\mathbf{f}^{kd}(\cdot) = 0$, which is again solved for the variables $\mathbf{x}_i^{kd}$ for each data point separately:

$$\mathbf{f}^{kd}(\hat{\mathbf{x}}_i^{ko}, \hat{\mathbf{y}}_i^{ko}, \mathbf{x}_i^{kd}, \tilde{\mathbf{y}}_i^{kd}) = 0. \tag{20}$$

Depending on the initial values used, the solution of the nonlinear optimization problems in Eq. (19) and the nonlinear equations in Eq. (20) may require many iterations and convergence may be difficult to achieve. Therefore, we propose to use the values of successful solutions of already processed data points as a look-up table for the initialization of the remaining data points. Thereby, convergence can be improved and accelerated.

Finally, the parameters are estimated from all data points by solving the constrained parameter estimation problem

$$\min_{\mathbf{x}_i^{ku},\mathbf{x}_i^u,\mathbf{z}_i,\mathbf{p}} \frac{1}{2}\sum_{i=1}^{n}(\mathbf{z}_i - \tilde{\mathbf{z}}_i)^T W^{ku,u}(\mathbf{z}_i - \tilde{\mathbf{z}}_i) + \lambda R(\mathbf{p}) \tag{21}$$

$$s.t. \quad \mathbf{f}^{ku}(\mathbf{x}_i^{ku},\mathbf{z}_i) = 0$$

$$\mathbf{f}^u(\mathbf{x}_i^u,\mathbf{z}_i,\mathbf{p}) = 0$$

$$\mathbf{g}(\mathbf{x}_i^u,\mathbf{z}_i,\mathbf{p}) \le 0$$

$$\mathbf{x}_l^{ku} \le \mathbf{x}_i^{ku} \le \mathbf{x}_u^{ku}, \quad \mathbf{x}_l^u \le \mathbf{x}_i^u \le \mathbf{x}_u^u,$$

$$\mathbf{z}_l \le \mathbf{z}_i \le \mathbf{z}_u, \quad \mathbf{p}_l \le \mathbf{p} \le \mathbf{p}_u$$

$$\forall \quad i = 1,\ldots,n,$$

which may contain inequality constraints $\mathbf{g}(\cdot)$ to prevent saturation of the neurons of the neural networks. The variable nomenclature was changed to ease notation. The variables $\tilde{\mathbf{z}}$ denote those variables of $\mathbf{f}^{ku}(\cdot)$ and $\mathbf{f}^u(\cdot)$ that are either measured, i.e. $\tilde{\mathbf{y}}_i^{kd}$, $\tilde{\mathbf{y}}_i^{ku}$ and $\tilde{\mathbf{y}}_i^u$, or have already been estimated, i.e. $\hat{\mathbf{x}}_i^{ko}$, $\hat{\mathbf{y}}_i^{ko}$, or $\hat{\mathbf{x}}_i^{kd}$.

In the incremental approach, errors in the estimated variables are propagated from the overdetermined part to the determined part, and from there to the underdetermined part. Therefore, the estimated variables and parameters will usually be close, but not identical, to the statistically sound estimates obtained by solving the simultaneous parameters estimation problem (Eq. 14). Thus, the simultaneous approach is finally used to efficiently remove the bias in the estimates obtained by the incremental approach.

Further details on the generalized incremental identification approach as well as algorithms to assess the hybrid model validity and extrapolability are presented in [13]. Furthermore, a prototypical implementation of a hybrid modeling environment is presented to demonstrate the applicability of the generalized incremental identification approach in several case studies.

6 Conclusions

The dynamic behavior of process systems is typically not sufficiently well understood to facilitate the formulation of a unique and unambiguous model structure to properly represents the overall process. Therefore, either linear or nonlinear model identification is often the method of choice to derive an empirical model from plant experiments in industrial practice. Such an approach completely neglects the prior knowledge which is always available. Hybrid modeling facilitates capturing the available knowledge into a mathematical process model and to complement the missing model features by a data-driven approach. Hybrid modeling, though attractive from an industrial perspective, did not yet get the necessary attention by systems engineering research which is required to devise a powerful methodology which can easily be followed in practice. Consequently, commercially available tools for the development and analysis of hybrid models are largely lacking. This contribution introduced a novel approach which exploits the natural structure in process systems models in the modeling strategy. The conceptually simple idea has been validated and illustrated by means of a few examples. These studies have identified the existing gaps which need to be filled to provide readily useful systematic modeling methods and software tools. Most notably, hybrid model structure development, synthesis of appropriate empirical model structures, identifiability and optimal experimental design techniques, more powerful algorithms for nonlinear parameter identification as well as analysis tools assessing the properties and range of validity of the resulting models are largely open issues, which need much more attention in future research work in systems and control. Obviously, the ideas presented in this paper, do carry over to other domains including mechatronics, automotive and aerospace systems as well as energy process systems.

References

[1] Agarwal, M.: Combining neural and conventional paradigms for modelling, prediction and control. Int. J. Syst. Sci. **28**, 65–81 (1997)
[2] Bard, Y.: Nonlinear Parameter Estimation. Academic Press, New York (1974)
[3] Bardow, A., Marquardt, W.: Identification Methods for Reaction Kinetics and Transport. In: Floudas, C.A., Pardalos, P.M. (eds.), Encyclopedia of Optimization, 2nd ed., Springer US, 1549–1556 (2009)
[4] Bonvin, D., Rippin, D.W.T.: Target factor analysis for the identification of stoichiometric models. Chem. Eng. Sci. **45**, 3417–3426 (1990)
[5] Brendel, M., Mhamdi, A., Bonvin, D., Marquardt, W.: An incremental approach for the identification of reaction kinetics. ADCHEM 2003, 177–182 (2003)
[6] Brendel, M., Marquardt, W.: Experimental design for the identification of hybrid reaction models from transient data. Chem. Eng. J. **141**, 264–277 (2009)

[7] Chang, J.S., Hung, B.C.: Optimization of batch polymerization reactors using neural-network rate function models. Ind. Eng. Chem. Res. **11**, 2716–2727 (2002)

[8] Dulmage, A.L., Mendelsohn, N.S.: Two algorithms for bipartite graphs. SIAM Journal **11**, 183–194 (1963)

[9] Froment, G.F., Bischoff, K.B.: Chemical Reactor Analysis and Design. John Wiley and Sons, New York. (1990)

[10] Hansen, P.C.: Rank-deficient and Discrete Ill-posed Problems. SIAM, Philadelphia (1998)

[11] Kahrs, O., Marquardt, W.: Incremental identification of hybrid process models. Comput. Chem. Eng. **32**, 694–705 (2007)

[12] Kahrs, O., Marquardt, W.: The validity domain of hybrid models and its application in process optimization. Chem. Eng. Prog. **46**, 1041–1242 (2007)

[13] Kahrs, O.: Semi-Empirical Modeling of Process Systems. PhD Thesis, RWTH Aachen University, Germany (2009)

[14] Van Lith, P.F., Betlem, B.H.L., Roffel, B.: A structured modeling approach for dynamic hybrid fuzzy first-principles models. J. Proc. Cont. **12**, 605–615 (2002)

[15] Marquardt, W.: Towards a process modeling methodology. In: Berber, R. (ed) Methods of Model-based Control. NATO-Asi Series, Kluwer, The Netherlands, 3–41 (1995)

[16] Marquardt, W.: Model-based Experimental Analysis of Kinetic Phenomena in Multi-phase Reactive Systems. Trans IChemE, Part A, Chemical Engineering Research and Design, **83**, 561–573 (2005)

[17] Michalik, C., Chachuat, B., Marquardt, W.: Incremental Global Parameter Estimation in Dynamical Systems. submitted (2009)

[18] Olivera, R.: Combining first principles modelling and artificial neural networks: a general framework. Comp. Chem. Eng. **28**, 755–766 (2004)

[19] Pantelides, C.C., Urban, Z.E.: Process Modelling Technology: A Critical Review of Recent Developments. In: Floudas, C.A., Agarwal, R. (eds.) Proc. Int. Conf. on Foundations of Process Design, FOCAPD 2004, 69–83 (2004)

[20] Psichogios, D.C., Ungar, L.H.:A hybrid neural network - first principles approach to process modeling. AIChE J. **38**, 1499–1511 (1992)

[21] Ruppen, D., Bonvin, D., Rippin, D.W.T.: Implementation of adaptive optimal operation for a semi-batch reactor. Comp. Chem. Eng. **22**, 185–199 (1997)

[22] Tholodur, A., Ramirez, W.F.: Optimization of fed batch bioreactors using neural net parameter function models. Biotechnol. Prog. **12**, 302–309 (1996)

[23] Yeow, Y.L., Wickramasinghe, S.R., Han, B., Leong, Y.K.: A new method of processing the time-concentration data of reaction kinetics. Chem. Eng. Sci. **58**, 3601–3610 (2003)

Front Controllability in Two-Phase Porous Media Flow

Jan Dirk Jansen, Jorn F.M. Van Doren, Mohsen Heidary-Fyrozjaee and Yannis C. Yortsos

Abstract The propagation of the front (i.e. the interface) between two immiscible fluids flowing through a porous medium is governed by convection, i.e. by the fluid velocities at the front, which in turn are governed by the pressure gradient over the domain. We investigated a special case of immiscible two-phase flow that can be described as potential flow, in which case the front is sharp and can be traced with a simple Lagrangian formulation. We analyzed the controllability of the pressure field, the velocity field and the front position, for an input in the form of slowly time-varying boundary conditions. In the example considered in this paper of order one equivalent aspect ratio, controllability of the pressures and velocities at the front to any significant level of detail is only possible to a very limited extent. Moreover, the controllability reduces with increasing distance of the front from the wells. The same conclusion holds for the local controllability of the front position, i.e. of changes in the front position, because they are completely governed by the velocities. Aspect ratios much lower than one (for instance resulting from strongly anisotropic permeabilities) or geological heterogeneities (for instance in the form of high-permeable streaks) are an essential pre-requisite to be able to significantly influence subsurface fluid flow through manipulation of well rates.

Jan Dirk Jansen
Delft University of Technology, Department of Geotechnology, PO box 5048, 2600 GA, Delft, The Netherlands, and Shell International E&P, Kessler Park 1, 2288 GS Rijswijk, The Netherlands, e-mail: j.d.jansen@tudelft.nl

Jorn F.M. Van Doren
Delft Center for Systems and Control, Delft University of Technology, Mekelweg 2, 2628 CD Delft, The Netherlands, e-mail: j.f.m.vandoren@tudelft.nl and Shell International E&P, Kessler Park 1, 2288 GS Rijswijk, The Netherlands, e-mail: jorn.vandoren@shell.com

Mohsen Heidary-Fyrozjaee
University of Southern California, Viterbi School of Engineering, 3650 McClintock Ave, Los Angeles, CA 90089 USA, e-mail: heidarif@usc.edu

Yannis C. Yortsos
University of Southern California, Viterbi School of Engineering, 3650 McClintock Ave, Los Angeles, CA 90089 USA, e-mail: yortsos@usc.edu

P.M.J. Van den Hof et al. (eds.), *Model-Based Control: Bridging Rigorous Theory and Advanced Technology*, DOI: 10.1007/978-1-4419-0895-7_12,
© Springer Science + Business Media, LLC 2009

1 Introduction

Fyrozjaee and Yortsos [5] investigated the scope to control the front between two fluids in a porous medium through prescribing the flow rates in spatially distributed sources and sinks. By disregarding fluid-fluid interactions and assuming identical fluid properties they obtained a potential flow description which allows for (semi) analytical solutions. In particular they developed an inverse method to compute the necessary source and sink strengths to achieve a predefined front movement in time and space. They showed that the inverse problem can be expressed as an integral equation allowing for analytical expressions for the kernels (Green's functions), which can be used to numerically compute the unknown source and sink strengths. They found that the numerical problem becomes increasingly ill-conditioned with increasing distance between the front and the control points (i.e. the sources and sinks) and requires some form of regularization to achieve meaningful results. Moreover, these results are approximate in the sense that the computed source/sink strengths do not exactly produce the desired front movement when used in a forward simulation. Zandvliet et al. [15] analyzed the controllability of the pressure field of single-phase porous media flow, and Van Doren [12] extended this analysis to two-phase flow. Here we will analyze in particular the deterioration of the ability to control the fluid front position in two-phase potential flow with increasing distance from the control points.

2 Front dynamics

Two-phase flow of two immiscible phases (e.g. oil and water) through a porous medium is governed by two coupled partial differential equations: a parabolic (diffusion) equation for the pressures, which in the limit of incompressibility reduces to an elliptic equation, and a parabolic-hyperbolic (convection) equation for one of the phase saturations[1], which in the absence of capillary forces reduces to a strictly (nonlinear) hyperbolic equation; see e.g. [1]. Here we consider the special case of two immiscible incompressible fluids with identical viscosities in a subsurface reservoir consisting of a homogeneous anisotropic porous medium, and we neglect gravity and capillary effects. In that case the saturations are either zero or one, i.e. the front between the two phases is sharp and its propagation is governed by pure convection driven by the local fluid velocities. Moreover, we assume that the pressure is not varying extremely fast, so that it is governed by a quasi-steady-state elliptic pressure equation, with time-dependent boundary conditions. We choose a two-dimensional description with spatial coordinates x' and y' aligned with the principal permeabilities such that the permeability[2] tensor is diagonal with permeabilities k_x

---

[1] Oil and water saturations are defined as the (dimensionless) fractions of the pore space occupied by the corresponding phase.

[2] Permeability is a measure of the ease which which fluids flow through a porous medium. In SI units it is expressed in m^2

and k_y. Through combination of the mass balance equation, and an empirical relationship between pressure drop and fluid velocity, known as Darcy's law, we obtain the following pressure equation (see e.g. Aziz and Settari [1]):

$$\frac{\partial}{\partial x'}\left(\frac{hk_x}{\mu}\frac{\partial p'(x',y',t')}{\partial x'}\right) + \frac{\partial}{\partial y'}\left(\frac{hk_y}{\mu}\frac{\partial p'(x',y',t')}{\partial y'}\right) = 0, \qquad (1)$$

where p' is pressure, t' is time, h reservoir height and μ fluid viscosity. We consider a horizontal rectangular reservoir with linear dimensions

$$0 \le x' \le L_x, \qquad (2)$$
$$0 \le y' \le L_y, \qquad (3)$$

and with boundary conditions

$$x' = 0: \quad \frac{\partial p'}{\partial x'} = -\frac{\mu}{hk_x}q'(y',t') \qquad (4)$$
$$x' = L_x: \quad p' = p'_0 \qquad (5)$$
$$y' = 0: \quad \frac{\partial p'}{\partial y'} = 0 \qquad (6)$$
$$y' = L_x: \quad \frac{\partial p'}{\partial y'} = 0, \qquad (7)$$

where q' is flow rate per unit length, and where we have dropped the dependence of p' on space and time from the notation for clarity. The boundary conditions represent a horizontal injection well with a prescribed flow rate distribution $q'(y',t')$ at the West side of the reservoir, a horizontal production well with a prescribed bottom hole pressure p'_0 at the East side, and no-flow conditions at the North and South sides respectively; see Figure 1. This configuration is an idealized representation of a flat reservoir with a 'smart' horizontal injector equipped with a large number of rate-controlled inflow control valves, and a conventional horizontal producer, and was inspired by the examples treated in Brouwer and Jansen [3]. We assume that the wells operate under voidance replacement conditions with a total field rate Q. We introduce the dimensionless variables

$$x = \frac{x'}{L_x}, \qquad (8)$$
$$y = \frac{y'}{L_y}, \qquad (9)$$
$$R_L = \frac{L_x}{L_y}\sqrt{\frac{k_y}{k_x}}, \qquad (10)$$

$$t = \frac{t'Q}{hL_xL_y\phi}, \tag{11}$$

$$p = \frac{(p' - p_0')hk_xL_y}{Q\mu L_x}, \tag{12}$$

$$q = \frac{q'L_y}{Q}, \tag{13}$$

$$v_x = \frac{v_x'hL_y}{Q}, \tag{14}$$

$$v_y = \frac{v_y'hL_x}{QR_L^2}, \tag{15}$$

where ϕ is porosity and R_L is an equivalent aspect ratio that was used earlier in [5]. Note that $R_L = 0$ represents non-communicating layers (zero y-permeability), or a 'tall' reservoir, while the other extreme $R_L \gg 1$ is the limit of transverse equilibrium or of a 'thin' reservoir [13]. Also note that at scaled time $t = 1$ the injected volume Qt' just equals the pore volume $hL_xL_y\phi$ of the reservoir. With the aid of these dimensionless variables we can rewrite equation (1) and boundary conditions (4) to (7) in dimensionless form as

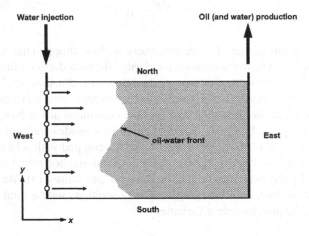

Fig. 1 Top view of a flat rectangular reservoir with a 'smart' horizontal injector equipped with inflow control valves, represented as a row of point sources, at the Western boundary, and a horizontal producer, represented as a line sink, at the Eastern boundary.

$$\frac{\partial^2 p}{\partial x^2} + R_L^2 \frac{\partial^2 p}{\partial y^2} = 0, \tag{16}$$

$$x = 0 \ : \quad \frac{\partial p}{\partial x} = -q(y,t) , \tag{17}$$

$$x = 1 \ : \quad p = 0 , \tag{18}$$

$$y = 0 \ : \quad \frac{\partial p}{\partial y} = 0 , \tag{19}$$

$$y = 1 \ : \quad \frac{\partial p}{\partial y} = 0 . \tag{20}$$

Following [5], we consider a sharp time-varying front between the two fluids that can be characterized with the expression

$$F\left(x',y',t'\right) = 0. \tag{21}$$

Because the front moves along with the fluid, the convective derivative is equal to zero, i.e.,

$$\frac{dF}{dt'} = \frac{\partial F}{\partial t'} + \frac{\partial F}{\partial x'}\frac{\partial x'}{\partial t'} + \frac{\partial F}{\partial y'}\frac{\partial y'}{\partial t'} = 0. \tag{22}$$

Introducing the dimensionless fluid velocities[3]

$$v_x = \frac{\partial x}{\partial t}, \tag{23}$$

$$v_y = \frac{1}{R_L^2}\frac{\partial y}{\partial t}, \tag{24}$$

we can rewrite equation (22) in dimensionless form as

$$\frac{\partial F}{\partial t} + v_x \frac{\partial F}{\partial x} + R_L^2 v_y \frac{\partial F}{\partial y} = 0. \tag{25}$$

Depending on the shape of the front it may be possible to replace the implicit equation (21) with an explicit one. In particular we consider situations where it is possible to write

$$F(x,y,t) = x - f(y,t), \tag{26}$$

such that equations (21) and (25) can be rewritten as

$$x = f(y,t), \tag{27}$$

and

$$\frac{\partial f}{\partial t} = v_x - R_L^2 v_y \frac{\partial f}{\partial y}, \tag{28}$$

---

[3] Expressed in unscaled variables the front moves with a true velocity instead of with the Darcy velocity, with components defined as $u_x' = \frac{v_x'}{\phi} = \frac{\partial x'}{\partial t'}$ and $u_y' = \frac{v_y'}{\phi} = \frac{\partial y'}{\partial t'}$.

which implies that we can also write

$$\frac{\partial x(y,t)}{\partial t} = v_x(x,y,t) - R_L^2 v_y(x,y,t) \frac{\partial x(y,t)}{\partial y}, \tag{29}$$

where we have re-introduced the dependence of the variables on space and time into the notation. Equation (29) is a nonlinear first-order hyperbolic equation that can be interpreted as a nonlinear convection equation for the x coordinates of the front, i.e. the $x(y,t)$ coordinates for given values of y and t. In the remainder of the paper, we will consider the solution of the problem for the special case $R_L = 1$.

3 Analytical solution

Fyrozjaee and Yortsos [5] gave analytical expressions in terms of infinite series for the pressure and velocity fields resulting from boundary conditions (17) to (20), which have been reproduced in the Appendix. They approximated the line sources at the Western and Eastern boundaries with a finite number m of rate-controlled point-sources $q_i, i = 1, \ldots, m$ and a full-length constant pressure source respectively. Figure 2 displays an example of the pressure and velocity fields resulting from a line source with unit flow rate defined as

$$q(y,t) = 0.5 + y + 0.8\cos(3\pi y), \tag{30}$$

and computed with the aid of expressions (49), (55) and (56).

4 Numerical approximation

To obtain a numerical approximation of the front propagation equation (29) we could attempt to use a finite difference representation for the convective term $v_y \frac{\partial x}{\partial y}$ and express the problem in terms of state variables, $x_i(y_i,t)$, $i = 1, \ldots, n$, representing the fluid front position at n discrete values y_i as a function of t. However, such an Eulerian description of the purely convective front movement is prone to numerical inaccuracies. It is easier to use a Lagrangian approach and simply track the movement in x and y directions of a number of 'fluid particles' at the front using equations (23) and (24). With the aid of the approximation

$$\frac{\partial x}{\partial t} \approx \frac{\Delta x}{\Delta t} = \frac{x_{k+1} - x_k}{\Delta t}, \tag{31}$$

we can write

$$x_{k+1} \;=\; x_k + v_x\left(x_k, y_k, t\right)\Delta t, \tag{32}$$

$$y_{k+1} \;=\; y_k + R_L^2 v_y\left(x_k, y_k, t\right)\Delta t, \tag{33}$$

where the subscript k indicates discrete time according to $t_k = k\Delta t$. Restricting ourselves to the case where $R_L = 1$, we can use the analytical expressions for v_x and v_y as given in Equations (55) and (56) in the Appendix to define a discrete-time nonlinear system equation

$$\mathbf{z}_{k+1} = \mathbf{z}_k + \mathbf{G}_z\left(\boldsymbol{\eta}, \mathbf{z}_k\right)\mathbf{q}_k \Delta t, \tag{34}$$

where $\boldsymbol{\eta} \in \mathbb{R}^m$ is the vector of point source locations, with m the number of sources, $\mathbf{z} \in \mathbb{R}^{2n}$ is a state vector given by

$$\mathbf{z} = \begin{bmatrix} \mathbf{x}^T & \mathbf{y}^T \end{bmatrix}^T = \begin{bmatrix} x_1 & x_2 & \cdots & x_n & y_1 & y_2 & \cdots & y_n \end{bmatrix}^T, \tag{35}$$

with n a finite number of points at the front location, and $\mathbf{q} \in \mathbb{R}^m$ is the input vector of point source strengths. The matrix $\mathbf{G}_z(\boldsymbol{\eta}, \mathbf{z}) \in \mathbb{R}^{2n \times m}$ is defined as

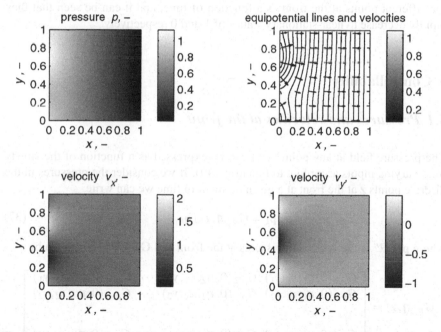

Fig. 2 Pressure and velocity fields resulting from 50 constant-rate point sources approximating a line source with a strength according to equation (30) at the Western boundary, and a full-length constant-pressure line source at the Eastern boundary, for $R_L = 1$. The values have been computed on a 50 by 50 grid, using 200 terms in the trigoniometric summations.

$$\mathbf{G}_z(\boldsymbol{\eta},\mathbf{z}) = \begin{bmatrix} G_{x,1}(0,\eta_1;x_1,y_1) & G_{x,2}(0,\eta_2;x_1,y_1) & \cdots & G_{x,m}(0,\eta_m;x_1,y_1) \\ G_{x,1}(0,\eta_1;x_2,y_2) & G_{x,2}(0,\eta_2;x_2,y_2) & \cdots & G_{x,m}(0,\eta_m;x_2,y_2) \\ \vdots & \vdots & \ddots & \vdots \\ G_{x,1}(0,\eta_1;x_n,y_n) & G_{x,2}(0,\eta_2;x_n,y_n) & \cdots & G_{x,m}(0,\eta_m;x_n,y_n) \\ G_{y,1}(0,\eta_1;x_1,y_1) & G_{y,2}(0,\eta_2;x_1,y_1) & \cdots & G_{y,m}(0,\eta_m;x_1,y_1) \\ G_{y,1}(0,\eta_1;x_2,y_2) & G_{y,2}(0,\eta_2;x_2,y_2) & \cdots & G_{y,m}(0,\eta_m;x_2,y_2) \\ \vdots & \vdots & \ddots & \vdots \\ G_{y,1}(0,\eta_1;x_n,y_n) & G_{y,2}(0,\eta_2;x_n,y_n) & \cdots & G_{y,m}(0,\eta_m;x_n,y_n) \end{bmatrix},$$

$$(36)$$

with elements $G_{x,j}$ and $G_{y,j}$, $j = 1,\ldots,m$ as defined in Equations (57) and (58) in the Appendix. Starting from initial condition $\mathbf{z}_0 = \begin{bmatrix} \mathbf{0}^T & \boldsymbol{\eta}^T \end{bmatrix}^T$ we can now use equation (34) to track the front position in time[4]. The top row of Figure 3 displays the front positions resulting from a line source corresponding to equation (30) at different moments in time as computed with equation (34). It can be seen that the point sources only have an effect on the shape of the front during early times (top left figure) whereas thereafter the shape of the front remains nearly unaltered (top right figure). The bottom row of Figure 3 displays the corresponding x- and y-velocities for different points at the front as a function of time, and it can be seen that they rapidly decrease to near-stationary values of 1 and 0 respectively.

5 Controllability

5.1 Pressures and velocities at the front

The pressure field at any point (x,y) can be expressed as a function of the slowly time-varying inputs according to Equation (49). If we consider the pressures in the discrete points $\mathbf{z}$ at the front at a given moment in time we can write

$$\mathbf{p}_k = \mathbf{G}_p(\boldsymbol{\eta},\mathbf{z}_k)\mathbf{q}_k, \qquad (37)$$

where $\mathbf{p} \in \mathbb{R}^n$ is the vector of pressures at the front and $\mathbf{G}_p \in \mathbb{R}^{n\times m}$ is given by

$$\mathbf{G}_p(\boldsymbol{\eta},\mathbf{z}) = \begin{bmatrix} G_{p,1}(0,\eta_1;x_1,y_1) & G_{p,2}(0,\eta_2;x_1,y_1) & \cdots & G_{p,m}(0,\eta_m;x_1,y_1) \\ G_{p,1}(0,\eta_1;x_2,y_2) & G_{p,2}(0,\eta_2;x_2,y_2) & \cdots & G_{p,m}(0,\eta_m;x_2,y_2) \\ \vdots & \vdots & \ddots & \vdots \\ G_{p,1}(0,\eta_1;x_n,y_n) & G_{p,2}(0,\eta_2;x_n,y_n) & \cdots & G_{p,m}(0,\eta_m;x_n,y_n) \end{bmatrix},$$

$$(38)$$

---

[4] In practice, we took the initial condition for the x-coordinates at a distance $L_y/(2m)$ away from the Western boundary, where L_y/m is the distance between the point sources, to reduce errors resulting from replacing the line source with a finite number of point sources.

with elements $G_{p,j}$, $j = 1, \ldots, m$ as defined in Equation (50). Because of the elliptic nature of the pressure equation transients do not occur and the n elements of pressure vector $\mathbf{p}_k$ at an arbitrary time k are only a function of the m elements of input vector $\mathbf{q}_k$ at the same moment. The pressure vector $\mathbf{p}_k$ is therefore completely controllable if

$$\text{rank}\left(\mathbf{G}_p\left(\boldsymbol{\eta}, \mathbf{z}_k\right)\right) = n, \tag{39}$$

which implies that we should have $n \leq m$, i.e. $\mathbf{G}_p(\boldsymbol{\eta}, \mathbf{z}_k)$ should be square or flat, and, in addition, should have independent rows. Controllability [5] of the pressure vector $\mathbf{p}_k$ implies that it is possible to achieve any desired value for its elements by a suitable combination of inputs $\mathbf{q}_k$. However, this qualitative controllability criterion

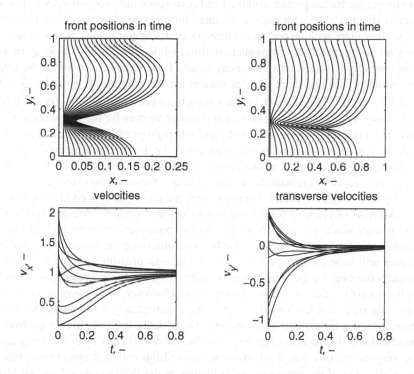

Fig. 3 Front positions and velocities as function of time. Top left: front positions for $0 \leqslant t \leqslant 0.16$ displayed at $\Delta t = 0.01$ with time increasing from left to right. Top right: front positions for $0 \leqslant t \leqslant 0.8$ displayed at $\Delta t = 0.05$ with time increasing from left to right. Bottom left: Velocities in x-direction for 10 of the 50 points at the front. Bottom right: Velocities in y-direction for the same points. All values have been computed with $\Delta t = 0.001$ for 50 points at the front, using 200 terms in the trigoniometric summations.

---

[5] For dynamic systems the necessary and sufficient conditions to reach a state $\mathbf{x}_{\underline{k}} = \check{\mathbf{x}}$ given the initial state $\mathbf{x}_0 = \mathbf{0}$ are often referred to as *reachability* conditions. The conditions for the reverse situation, i.e. to reach a state $\mathbf{x}_{\underline{k}} = \mathbf{0}$ given the initial state $\mathbf{x}_0 = \hat{\mathbf{x}}$ are then referred to as the controllability conditions. For this simple steady-state case the difference is irrelevant.

implies that magnitude of the input vector elements is unlimited, which is clearly not realistic. A more meaningful, quantitative, measure of controllability is obtained by considering the singular value decomposition (SVD) of $\mathbf{G}_p$:

$$\mathbf{G}_p(\boldsymbol{\eta},\mathbf{z}_k) = \mathbf{U}_k \boldsymbol{\Sigma}_k \mathbf{V}_k^T, \tag{40}$$

where, as usual, $\mathbf{U}$ is an $n \times n$ matrix of left singular values of $\mathbf{G}_p$, $\mathbf{V}$ an $m \times m$ matrix of right-singular vectors and $\boldsymbol{\Sigma}$ an $n \times m$ matrix with the non-zero singular values on its main diagonal. A similar analysis can be made of the controllability of the velocities $\mathbf{v}_x$ and $\mathbf{v}_y$ at the front using matrices $\mathbf{G}_x(\boldsymbol{\eta},\mathbf{z}) \in \mathbb{R}^{n \times m}$ and $\mathbf{G}_y(\boldsymbol{\eta},\mathbf{z}) \in \mathbb{R}^{n \times m}$ with elements $G_{x,j}$ and $G_{y,j}$ as defined in Equations (57) and (58) respectively. Figure 4 (top-left) displays the magnitude of the normalized singular values for three moments in time for the pressures at the front corresponding to equation (30). The three sets of singular values for each of the three moments in time have been normalized by scaling them with respect to the three largest singular values. It can be seen that they all decay rapidly at any moment in time, while, in addition, the decay becomes progressively stronger with increasing time[6]. For example, at $t = 0.1$, the relative magnitude of the first three singular vectors for the pressures is given by $\frac{\sigma_2}{\sigma_1} = 0.26$ and $\frac{\sigma_3}{\sigma_1} = 0.09$, while at $t = 0.7$ these values have been reduced to 0.08 and 0.05 respectively. The first three left and right singular vectors for the pressures at $t = 0.1$ have been displayed in the top-middle and top-right graphs of Figure 4. The left singular vectors correspond to 'pressure patterns' (i.e. linear combinations of pressures at the front) that can be controlled through varying the amplitude of the 'input patterns' (i.e. linear combinations of sources at the Western boundary) as represented by the right singular vectors. Similar figures have been displayed in the middle and bottom rows of Figure 4 for the singular values and vectors corresponding to the x- and y-velocities at the front. Note that for the pressures (top row) the right and left singular vectors are similar in shape because changing the strength of a particular source will have the strongest effect on the nearest pressure value at the front. The same is the case for the x-velocities (middle row), i.e. for the velocities at the front in the main direction of the front propagation. However, for the y-velocities (bottom row) the right and left singular vectors have completely different shapes because more complex input patterns are required to influence the velocities at the front perpendicular to the main front propagation direction. It is clear from the decay rates of the singular vectors that, in a homogeneous medium with unit aspect ratio, practical controllability of the pressures and velocities at the front to any significant level of detail is difficult right from the start and becomes more and more hopeless with increasing time. Note that controlling either the x- or the y-velocities of the front also influences the other velocity components, i.e. the y- or x-components respectively. The controllability of combined x- and y-velocities at the front can be quantified by taking the SVD of matrix $\mathbf{G}_z(\boldsymbol{\eta},\mathbf{z}) = [\ \mathbf{G}_x^T(\boldsymbol{\eta},\mathbf{z})\ \ \mathbf{G}_y^T(\boldsymbol{\eta},\mathbf{z})\ ]^T \in \mathbb{R}^{2n \times m}$.

---

[6] The 'kink' in the curves for $t = 0.4$ and $t = 0.7$ is caused by the inability to express numbers smaller than 10^{-16} in double precision, and the singular values with a magnitude below that value are therefore unreliable.

5.2 Position of the front

The position of the front $\mathbf{z}_K$ at a certain discrete time K can be expressed as a function of the time-varying inputs according to Equation (34). However, unlike for the pressure and the velocities at the front, we need to consider the entire input history as given by the terms $\mathbf{G}_z(\boldsymbol{\eta}, \mathbf{z}_k)\mathbf{q}_k$, $k = 1, \ldots, K - 1$ in Equation (34). Moreover, the dependency of the transfer functions $\mathbf{G}_z(\boldsymbol{\eta}, \mathbf{z}_k)$ on the front position implies that the front position is a nonlinear function of the inputs $\mathbf{q}_k$, $k = 1, \ldots, K - 1$. Van Doren [12] describes various approaches to analyze the controllability of saturation distributions in two-phase porous media flow. In particular he uses the concept of empirical Gramians, which were introduced by Lall et al. [8] and Hahn et al. [6]. These publications describe how empirical Gramians can be computed by performing a large number of simulations with different input sequences that should represent the 'typical' input behavior of the system along the entire input trajectory. However, if we restrict ourselves to an analysis of the local front position controllability, we can make use of the time-discretized approximation (34) for the front position $\mathbf{z}_{k+1} = [\ \mathbf{x}_{k+1}^T \ \ \mathbf{y}_{k+1}^T \ ]^T$. Because we have

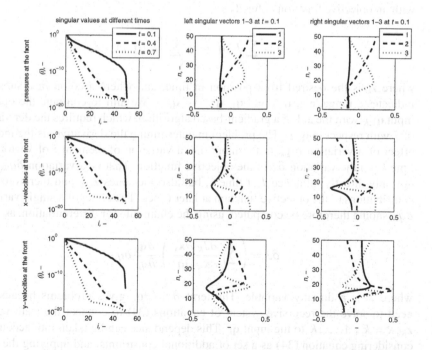

Fig. 4 : Left column: normalized singular values of matrices $\mathbf{G}_p$, $\mathbf{G}_x$ and $\mathbf{G}_y$, each of them at three moments in time. Middle column: first three left singular vectors (state patterns) of $\mathbf{G}_p$, $\mathbf{G}_x$ and $\mathbf{G}_y$ at $t = 0.1$. Right column: first three right singular vectors (input patterns) of $\mathbf{G}_p$, $\mathbf{G}_x$ and $\mathbf{G}_y$ at $t = 0.1$.

$$\mathbf{v}_{x,k} = \mathbf{G}_x(\boldsymbol{\eta}, \mathbf{z}_k)\mathbf{q}_k, \tag{41}$$

$$\mathbf{v}_{y,k} = \mathbf{G}_y(\boldsymbol{\eta}, \mathbf{z}_k)\mathbf{q}_k, \tag{42}$$

equation (34) is a linear function of the velocities $\mathbf{v}_{x,k}$ and $\mathbf{v}_{y,k}$. The change in front position at time k is thus completely governed by the velocities $\mathbf{v}_{x,k}$ and $\mathbf{v}_{y,k}$, and the controllability of these changes, i.e. the local controllability of the front positions, is therefore identical to the controllability of the velocities which was analyzed in Section 5.1 above.

6 Front control

Fyrozjaee and Yortsos [5] and Fyrozjaee [4] described a method to compute the required inputs $\mathbf{q}_{1:K}$ to achieve a desired front position based on a direct inversion of equation (29). Alternatively, we may pose the front control problem as a minimization problem

$$\min_{\mathbf{q}_{1:K}} J(\mathbf{q}_{1:K}), \tag{43}$$

with an objective function defined as

$$J(\mathbf{q}_{1:K}) = \sum_{k=1}^{K} J_k(\mathbf{q}_{1:k}) = \sum_{k=1}^{K}\left[\mathbf{z}_k(\mathbf{q}_{1:k}) - \check{\mathbf{z}}_k\right]^2, \tag{44}$$

where $\check{\mathbf{z}}_k$ is the desired front position in time, and where a colon in a subscript indicates a range, e.g. $\mathbf{q}_{1:K} = [\ \mathbf{q}_1 \ \mathbf{q}_2 \ \cdots \ \mathbf{q}_K\ ]$. We aim to compute the optimal input $\mathbf{q}_{1:K}$ with the aid of a gradient-based algorithm, which requires the derivatives of J with respect to $\mathbf{q}_{1:K}$. The problem in determining the derivatives is the indirect effect of a variation $\delta\mathbf{q}_{ik}$ in the input, i.e. a variation of element i of vector $\mathbf{q}$ at time k, on the variation δJ in the objective function. That is, a variation δq_{ik} does not only directly influence J at time k, but also the states $\mathbf{z}_{k:K}$ and therefore their contributions to the objective function at later times. The effect of a single variation δq_{ik} should therefore be computed, using the chain rule for differentiation, as

$$\delta J = \left(\sum_{\kappa=k}^{K} \frac{\partial J_\kappa}{\partial \mathbf{x}_\kappa}\frac{\partial \mathbf{x}_\kappa}{\partial \mathbf{q}_k}\right)\frac{\partial \mathbf{q}_k}{\partial q_{ik}}\delta q_{ik} \tag{45}$$

where κ is a dummy variable. The term $\partial x_\kappa/\partial\mathbf{q}_k$ gives problems because we need to solve the recursive system of equations (34) to connect the state vectors $\mathbf{z}_\kappa, \kappa = k+1, \ldots, K$ to the input $\mathbf{q}_k$. This dependence can be taken into account by considering equation (34) as a set of additional constraints, and applying the technique of Lagrange multipliers to solve the constrained optimization problem; see e.g. [10] for a general overview, and Sudaryanto and Yortsos [11], Brouwer and Jansen [3], Zandvliet et al. [14] and Jansen et al. [7] for applications to the control of multiphase flow in porous media. We will not further address the front control problem in this paper but note that the right-singular vectors of the velocity matrices $G_x(\eta, z)$, $G_y(\eta, z)$ and $G_z(\eta, z)$ appear to offer potential for effective regularization of the front control problem. Additional work is required to assess this scope.

7 Concluding remarks

In the homogeneous example considered in this paper, controllability of the pressures and velocities at the front to any significant level of detail is only possible to a very limited extent. Moreover, the controllability reduces with increasing distance of the front from the wells. The same conclusion holds for the local controllability of the front position, i.e. of changes in the front position, because they are completely governed by the velocities. This is in line with the findings of Fyrozjaee and Yortsos [5] who considered the same configuration and observed that an inverse approach to control the front position became increasingly ill-conditioned with increasing distance of the front from the wells. A similar conclusion was reached by Ramakrishnan [9]. We note that the existence of a small equivalent aspect ratio (either small transverse permeability, compared to streamwise, or a 'tall' reservoir) is an essential pre-requisite for significantly influencing subsurface fluid flow through manipulation of well rates. Moreover, the presence of heterogeneities in the form of high-permeable streaks and their correct orientation with respect to the wells may result in a considerable scope for front control; see Brouwer [2] and Fyrozjaee [4] for further discussions of these aspects.

Nomenclature

Symbol	Description	Dimensions	SI units
f	explicit function to specify front	-	-
F	implicit function to specify front	-	-
G	Green's function	-	-
$\mathbf{G}$	matrix of Green's functions	-	-
h	height	L	m
i	counter	-	-
j	counter	-	-
J	objective function	-	-
k	permeability	L^2	m^2
k	discrete time	-	-
L	length	L	m
m	number of source terms	-	-
n	summation term counter	-	-

continued on next page

Symbol	Description	Dimensions	SI units
n	number of discretization points at front	-	-
p	dimensionless pressure	-	-
p'	pressure	$L^{-1}mt^{-2}$	Pa
$\mathbf{p}$	dimensionless pressure vector	-	-
q	dimensionless flow rate per unit length	-	-
q'	flow rate per unit length	L^2t^{-1}	m^2/s
$\mathbf{q}$	dimensionless source vector (input vector)	-	-
Q	flow rate	L^3t^{-1}	m^3/s
R	ratio	-	-
t	dimensionless time	-	-
t'	time	t	s
u'	velocity	Lt^{-1}	m/s
v	dimensionless velocity	-	-
v'	Darcy velocity (superficial velocity)	Lt^{-1}	m/s
$\mathbf{v}$	vector of dimensionless front velocities	-	-
$\mathbf{V}$	system matrix	-	-
x	dimensionless spatial coordinate	-	-
x'	spatial coordinate	L	m
$\mathbf{x}$	vector of dimensionless x-coordinates at front	-	-
y	dimensionless spatial coordinate	-	-
y'	spatial coordinate	L	m
$\mathbf{y}$	vector of dimensionless y-coordinates at front	-	-
$\mathbf{z}$	state vector (dimensionless front position)	-	-
η	dimensionless source coordinate	-	-
$\boldsymbol{\eta}$	vector of dimensionless source coordinates	-	-
κ	dummy variable	-	-
λ	auxiliary variable	-	-
μ	viscosity	$L^{-1}mt^{-1}$	Pa s
σ	singular value	-	-
ϕ	porosity	-	-

Subscripts

d	discrete
k	discrete time
L	length
p	pressure
x	x direction
y	y direction
z	x and y directions

Superscripts

T	transpose

Appendix: Analytical expressions

Following Fyrozjaee and Yortsos [5], and taking $R_L = 1$, we specify an analytical expression for the pressure distribution corresponding to the prescribed-rate and prescribed-pressure line sources at the Western and Eastern reservoir boundaries:

$$p(x,y,t) = \int_0^1 G(0,\eta;x,y)\,q(\eta,t)\,d\eta, \tag{46}$$

where

$$\int_0^1 q(\eta,t)\,d\eta = 1 \tag{47}$$

at any time t, and where G is a Green's function defined as

$$G(0,\eta;x,y) = 1 - x + \sum_{n=1}^{\infty} \frac{2\sinh\left[\lambda_n(1-x)\right]\cos(\lambda_n y)}{\lambda_n \cosh \lambda_n}\cos\lambda_n\eta, \tag{48}$$

with $\lambda_n = n\pi$. The Green's function (48) represents the influence of a point source in $(x,y) = (0,\eta)$ on the pressure in an arbitrary point (x,y), and its derivation can be found in Fyrozjaee [4]. An approximation of the line source $q(\eta,t)$ in equation (46) with the aid of m point sources leads to q_j, $j = 1,\ldots,m$ leads to

$$p(x,y,t) = \sum_{j=1}^{m} G_{p,j}(0,\eta_j;x,y)\,q_j(\eta_j,t), \tag{49}$$

where

$$G_{p,j}(0,\eta_j;x,y) = 1 - x + \sum_{n=1}^{\infty} \frac{2\sinh\left[\lambda_n(1-x)\right]\cos(\lambda_n y)}{\lambda_n \cosh \lambda_n}\cos\lambda_n\eta_j, \tag{50}$$

with

$$\eta_j = \frac{j}{m} - \frac{1}{2m}, \tag{51}$$

and

$$\sum_{j=1}^{m} q_j(\eta_j,t) = 1. \tag{52}$$

Using the scaled version of Darcy's law

$$v_x = -\frac{\partial p}{\partial x}, \tag{53}$$

$$v_y = -\frac{\partial p}{\partial y}, \tag{54}$$

the velocity components corresponding to pressure distribution (49) are obtained as

$$v_x(x,y,t) = \sum_{j=1}^{m} G_{x,j}(0,\eta_j;x,y)\,q_j(\eta_j,t), \tag{55}$$

and

$$v_y(x,y,t) = \sum_{j=1}^{m} G_{y,j}(0,\eta_j;x,y)\,q_j(\eta_j,t), \tag{56}$$

where

$$G_{x,j}(0,\eta_j;x,y) = 1 + \sum_{n=1}^{\infty} \frac{2\cosh[\lambda_n(1-x)]\cos(\lambda_n y)}{\cosh\lambda_n}\cos\lambda_n\eta_j, \tag{57}$$

and

$$G_{y,j}(0,\eta_j;x,y) = \sum_{n=1}^{\infty} \frac{2\sinh[\lambda_n(1-x)]\sin(\lambda_n y)}{\cosh\lambda_n}\cos\lambda_n\eta_j. \tag{58}$$

References

[1] Aziz, K., Settari, A.: Petroleum Reservoir Simulation. Applied Science Publishers (1979)

[2] Brouwer, D.R.: Dynamic Water Flood Optimization With Smart Wells Using Optimal Control Theory. Ph. D. thesis, Delft University of Technology (2004)

[3] Brouwer, D.R., Jansen, J.D.: Dynamic optimization of waterflooding with smart wells using optimal control theory. SPE Journal (SPE 78278) 9(4), 391–402 (2004)

[4] Fyrozjaee, M.H.: Control of Displacement Fronts in Potential Flow using Flow-Rate Partitioning. Ph. D. thesis, University of Southern California (2008)

[5] Fyrozjaee, M.H., Y. C. Yortsos, Y.C.: Control of a displacement front in potential flow using flow rate partition. In: SPE Intelligent Energy Conference and Exhibition (SPE 99524), Amsterdam (2006)

[6] Hahn, J., Edgar, T.F., Marquardt. W.: Controllability and observability covariance matrices for the analysis and order reduction of stable nonlinear systems. Journal of Process Control 13(2), 115–127 (2003)

[7] Jansen, J.D., Bosgra, O.H., Van den Hof, P.M.J.: Model-based control of multiphase flow in subsurface oil reservoirs. Journal of Process Control 18(9), 846–855 (2008)

[8] Lall, S., Marsden, J.E., Glavaski, S.A.: A subspace approach to balanced truncation for model reduction of nonlinear control systems. International Journal of Robust Nonlinear Control 12(6), 519 (2002)

[9] Ramakrishnan, T.S.: On reservoir fluid-flow control with smart completions. SPE Production and Operations 22(1), 4–12 (2007)

[10] Stengel, R.F.: Stochastic Optimal Control: Theory and Application. John Wiley & Sons (1986)

[11] Sudaryanto, B., Yortsos, Y.C.: Opimization of fluid front dynamics in porous media flow using rate control. Physics of Fluids 12(7), 1656–1670 (2000)

[12] Van Doren, J.F.M.: Model Structure Analysis for Model-Based Operation of Petroleum Reservoirs. Ph. D. thesis, Delft University of Technology (in preparation)

[13] Yang, Z., Yortsos, Y. C., and Salin, D.: Asymptotic regimes of unstable miscible displacements in random porous media. Adv. Water Res., special anniversary issue **25**, 885–898 (2002)
[14] Zandvliet, M.J., Bosgra, O.H., Jansen, J.D., Van den Hof, P.M.J, Kraaijevanger, J.F.B.M.: Bang-bang control and singular arcs in reservoir flooding. Journal of Petroleum Science and Engineering **58**(1&2), 186–200 (2007)
[15] Zandvliet, M.J., Van Doren, J.F.M., Bosgra, O.H., Jansen, J.D., Van den Hof, P.M.J.: Controllability, observability and identifiability in single-phase porous media flow. Computational Geosciences **12**(4), 605–622 (2008)

Part IV
Appendix

PhD Supervision by Okko H. Bosgra

Delft University of Technology

1) W.J. Neaije. *Optimale uitgangsterugkoppeling als ontwerpmethode voor multivariabele regelsystemen,* (in Dutch), 1979.
2) G.A. Van Zee. *System identification for multivariable control,* 1981.
3) A.J.J. Van der Weiden. *The use of structural properties in linear multivariable control system design,* 1983.
4) B. Jonker. *A finite element dynamic analysis of flexible spatial mechanisms and manipulators,* 1988.
5) J.M.P. Van der Looij. *Dynamic modeling and control of coal fired fluidized bed boilers,* 1988.
6) M. Steinbuch. *Dynamic modelling and robust control of a wind energy conversion system,* 1989.
7) H. Aling. *Identification of closed loop systems: identifiability, recursive algorithms and application to a power plant,* 1990.
8) S. De Wolf. *Modelling, system identification and control of an evaporative continuous crystallizer,* 1990.
9) P.S.C. Heuberger. *On approximate system identification with system based orthonormal functions,* 1991.
10) R.J.P. Schrama. *Approximate identification and control design: with application to a mechanical system,* 1992.
11) F.G. De Boer. *Multivariable servo control of a hydraulic RRR-robot,* 1992.
12) H. Huisman. *Design and control of a class of multiphase series-resonant power converters,* 1992.
13) P. Schoen. *Dynamic modeling and control of integrated coal gasification combined cycle units,* 1993.
14) P.J.T. Venhovens. *Optimal control of vehicle suspensions,* 1994.
15) P.M.M. Bongers. *Modeling and identification of flexible wind turbines and a factorizational approach to robust control design,* 1994.

223

224

16) P.M.R. Wortelboer. *Frequency-weighted balanced reduction of closed-loop mechanical servo-systems: theory and tools*, 1994.
17) P.F. Lambrechts. *The application of robust control theory concepts to mechanical servo systems*, 1994.
18) D.K. De Vries. *Identification of model uncertainty for control design*, 1994.
19) R.G. Hakvoort. *System identification for robust process control. Nominal models and error bounds*, 1994.
20) R.A. Eek. *Control and dynamic modelling of industrial suspension crystallizers*, 1995.
21) G.W. Van der Linden. *High performance manipulator motion control Theory, tools and experimental implementation*, 1997.
22) J. Heintze. *Design and control of a hydraulically actuated industrial brick laying robot*, 1997.
23) G. Van Schothorst. *Modelling of long-stroke hydraulic servo-systems for flight simulator motion control and system design*, 1997.
24) J. Schuurmans. *Control of water levels in open channels*, 1997.
25) H.J.C. Gommeren. *Study of a closed circuit jet mill plant using on-line particle size measurements*, 1997.
26) D. De Roover. *Motion control of a wafer stage: A design approach for speeding up IC production*, 1997.
27) W.R. Pasterkamp. *The tyre as sensor to estimate friction*, 1997.
28) J.G.M. Dötsch. *Identification for control design of a compact disc mechanism*, 1998.
29) J. Do Livramento. *Dynamic modelling for analysis and design of bottle conveying systems in high-speed bottling lines*, 1998.
30) S.K. Advani. *The kinematic design of flight simulator motion bases*, 1998.
31) R.A. De Callafon. *Feedback oriented identification for enhanced and robust control: a fractional approach applied to a wafer stage*, 1998.
32) H. Smakman. *Functional integration of slip control with active suspension for improved lateral vehicle dynamics*, 2000.
33) W. He. *Dynamic simulations of molten-carbonate fuel-cell Systems*, 2000.
34) E.T. Van Donkelaar. *Model predictive control and system identification of industrial processes - A flexible linear parameterization approach*, 2000.
35) J.F. Kikstra. *Modelling design and control of a cogenerating nuclear gas turbine plant*, 2001.
36) M. Dettori. *LMI techniques for control - with application to a Compact Disc player mechanism*, 2001.
37) T.J. De Hoog. *Rational orthonormal bases and related transforms in linear system modeling*, 2001.
38) S.H. Koekebakker. *Model based control of a flight simulator motion system*, 2001.
39) R.L. Tousain. *Dynamic optimization in business-wide process control*, 2002.
40) D.P. Molenaar. *Cost-effective design and operation of variable speed wind turbines*, 2003.

41) D.H. Van Hessem. *Stochastic inequality constrained closed-loop model predictive control with application to chemical process operation*, 2004.

42) B.G. Dijkstra. *Iterative Learning Control, with applications to a wafer-stage*, 2004.

43) M. Stork. *Model-based optimization of the operation procedure of emulsification*, 2005.

44) J. Van den Berg. *Model reduction for dynamic real-time optimization of chemical processes*, 2005.

45) P.J. Van Overloop. *Model predictive control on open water systems*, 2006.

46) C.W.J. Hol. *Structured controller synthesis for mechanical servo-systems: algorithms, relaxations and optimality certificates*, 2006.

47) A.T.M.J. Soetjahjo. *Mathematical analysis of dynamic process models; index, inputs and interconnectivity*, 2006.

48) M.B. Groot Wassink. *Inkjet printhead performance enhancement by feedforward input design based on two-port modeling*, 2007.

49) L. Jabben. *Mechatronic design of a magnetically suspended rotating platform*, 2007.

50) M.J. Zandvliet. *Model-based lifecycle optimization of well locations and production settings in petroleum reservoirs*, 2008.

Wageningen University and Research Centre

1) W. Huisman. *Optimum cereal combine harvester operation by means of automatic machine and threshing speed control*, 1983.

Eindhoven University of Technology

1) P.H.M. Janssen. *On model parametrization and model structure selection for identification of MIMO-systems*, 1988.

2) P.M.J. Van den Hof. *On residual-based parametrization and identification of multivariable systems*, 1989.

3) A.G. De Jager. *Practical evaluation of robust control for a class of nonlinear mechanical dynamic systems*, 1992.

4) A.J.J. Van den Boom. *MIMO system identification for H_∞ robust control: a frequency domain approach with minimum error bounds*, 1993.

5) H.M. Falkus. *Parametric uncertainty in system identification*, 1994.

6) J.H.A. Ludlage. *Controllability analysis of industrial processes: towards the industrial application*, 1997.

7) S.G. Stan. *Optimization of the CD-ROM system towards higher data throughputs*, 1999.

226

8) A. Tyagunov. *High-performance model predictive control for process industry*, 2004.
9) D. Kostic. *Data-driven robot motion control design*, 2004.
10) P.W.J.M Nuij. *Higher order sinusoidal input describing functions: extending linear techniques towards non-linear systems analysis*, 2007.

Okko H. Bosgra,
Bibliographic Record

Journal Papers, Book Chapters and Book Reviews

[1] Backx, T., Bosgra, O.H., Marquardt, W.: Integration of model predictive control and optimization of processes. In: L. Biegler, A. Brambilla, C. Scali, G. Marchetti (eds.) Advanced Control of Chemical Processes 2000 (ADCHEM 2000), pp. 249–260. Elsevier (2001)

[2] Bongers, P., van Baars, G.E., Dijkstra, S., Bosgra, O.H.: Dynamic models for wind turbines. Delft University of Technology, MEMT-28, Delft, The Netherlands (1993)

[3] Bongers, P.M.M., Bosgra, O.H.: Normalized coprime factorizations for systems in generalized state-space form. IEEE Transactions on Automatic Control 38(2), 348–350 (1993)

[4] Bosgra, O.H.: On parametrizations for the minimal partial realization problem. Systems & Control Letters 3(4), 181–187 (1983)

[5] Bosgra, O.H.: Multivariable feedback control - analysis and design, Skogestad & Postlethwaite. IEEE Control Systems Magazine 27(1), 80–81 (2007)

[6] Bosgra, O.H., Engell, S., H., M.R.: Discussion on: Limitations on control system performance by K.J.Åstrom. European Journal of Control 6(1), 21–26 (2000)

[7] Bosgra, O.H., Lambrechts, P., Steinbuch, M.: μ-analysis and synthesis toolbox (μ-Tools) - a software review. Automatica 30, 733–735 (1994)

[8] Bosgra, O.H., Steinbuch, M., Smit, G.C.: Robust control of high performance mechanical servo-mechanisms. Journal A 31, 96 101 (1990)

[9] Bosgra, O.H., Tichem, M.: Research perspectives for the TU Delft mechatronics and microsystems research program. In: M. Tichem, J.F.L. Goosen, H.H. Langen, A.J.J. van der Weiden (eds.) Mechatronics and Microsystems, pp. 1–8. Delft University of Technology, Delft, The Netherlands (2003)

[10] Bosgra, O.H., van der Weiden, A.J.J.: Input-output invariants for linear multivariable systems. IEEE Transactions on Automatic Control 25(1), 20–36 (1980)

[11] Bosgra, O.H., van der Weiden, A.J.J.: Realizations in generalized state–space form for polynomial system matrices and the definitions of poles, zeros and decoupling zeros at infinity. International Journal of Control 33(3), 393–411 (1981)

[12] Douma, S.G., Van den Hof, P.M.J., Bosgra, O.H.: Controller tuning freedom under plant identification uncertainty: double Youla beats gap in robust stability. Automatica 39, 325–333 (2003)

[13] Draijer, W., Steinbuch, M., Bosgra, O.H.: Adaptive control of the radial servo system of a compact disc player. Journal A 31, 91–95 (1990)

[14] Draijer, W., Steinbuch, M., H., B.O.: Adaptive control of the radial servo system of a compact disc player. Automatica 28(3), 455–462 (1992)

[15] Eek, R.A., Bosgra, O.H.: Controllability of particulate processes in relation to the sensor characteristics. Powder technology 108, 137–146 (2000)

228

[16] Eek, R.A., Hoogenboezem, A.J., Bosgra, O.H.: Design issues related to the control of continuous crystallizers. Computers and Chemical Engineering 20(4), 427435 (1996)

[17] Eek, R.A., Pouw, H.A.A., Bosgra, O.H.: Design and experimental evaluation of stabilizing feedback controllers for continuous crystallizers. Powder Technology 82(1), 2135 (1995)

[18] de Gelder, E., van de Wal, M., Scherer, C.W., Hol, C.W.J., Bosgra, O.H.: Nominal and robust feedforward design with time domain constraints applied to a wafer stage. Journal of Dynamic Systems, Measurement and Control - Transactions of the ASME 128(2), 204–215 (2006)

[19] van Groos, P.J.M., Steinbuch, M., Bosgra, O.H.: Robust control of a compact disc player. Journal A 35, 57–62 (1994)

[20] Groot Wassink, M.B., van de Wal, M.M.J., Scherer, C.W., Bosgra, O.H.: LPV control of a wafer stage: beyond the theoretical solution. Control Engineering Practice 13(2), 231–245 (2005)

[21] van Hessem, D.H., Bosgra, O.H.: Stochastic closed-loop model predictive control of continuous nonlinear chemical processes. Journal of Process Control 16(3), 225–241 (2006)

[22] Heuberger, P.S.C., Van den Hof, P.M.J., Bosgra, O.H.: A generalized orthonormal basis for linear dynamical systems. IEEE Transactions on Automatic Control 40(3), 451465 (1995)

[23] Van den Hof, P.M.J., Schrama, R.J.P., Callafon, R.A.d., Bosgra, O.H.: Identification of normalized coprime plant factors from closed loop experimental data. European Journal of Control 1(1), 6274 (1995)

[24] Hol, C.W.J., Scherer, C.W., van der Meché, E.v.d., Bosgra, O.H.: A nonlinear SDP approach to fixed-order controller synthesis and comparison with two other methods applied to an active suspension system. European Journal of Control 9(1), 13–28 (2003)

[25] Jager, J., Kramer, H.J.M., de Jong E. J.: Control of industrial crystallizers. Powder Technology 69(1), 11–20 (1992)

[26] Jansen, J.D., Bosgra, O.H., Van den Hof, P.M.J.: Model-based control of multiphase flow in subsurface oil reservoirs. Journal of Process Control 18, 864–855 (2008)

[27] Leskens, M., van Kessel, L.B.M.v., Bosgra, O.H.: Model Predictive Control as a tool for improving the process operation of MSW combustion plants. Waste Management 25(8), 788–798 (2005)

[28] van der Meulen, S.H., Tousain, R., Bosgra, O.H.: Fixed structure feedforward controller design exploiting iterative trials: Application to a wafer stage and a desktop printer. Journal of Dynamic Systems, Measurement and Control - Transactions of the ASME 130(5), 051,006 (2008)

[29] Molenaar, D.P., Bosgra, O.H., Hoeijmakers, M.J.: Time-domain identification of synchronous generator transfer functions. Journal of Solar Energy Engineering - Transactions of the ASME 124, 419–426 (2002)

[30] Nuij, P., Steinbuch, M., Bosgra, O.H.: Experimental characterization of the stick/sliding transition in a precision mechanical system using the third order sinusoidal input describing function. Mechatronics 18, 100–120 (2008)

[31] Nuij, P., Steinbuch, M., Bosgra, O.H.: Measuring the higher order sinusoidal input describing functions of a non-linear plant operating in feedback. Control Engineering Practice 16(1), 101–113 (2008)

[32] Nuij, P.W.J.M., Bosgra, O.H., Steinbuch, M.: Higher-order sinusoidal input describing functions for the analysis of non-linear systems with harmonic responses. Mechanical Systems and Signal Processing 20(8), 1883–1904 (2006)

[33] Nuij, P.W.J.M., Steinbuch, M., Bosgra, O.H.: Experimental characterization of the stick/sliding transition in a precision mechanical system using the third order sinusoidal input describing function. Mechatronics 18(2), 100–110 (2008)

[34] Oomen, T., van de Wal, M., Bosgra, O.H.: Design framework for high-performance optimal sampled-data control with application to a wafer stage. International Journal of Control 80(6), 919–934 (2007)

[35] Oomen, T., van de Wijdeven, J., Bosgra, O.H.: Suppressing intersample behavior in iterative learning control. Automatica (2009). In Press

[36] de Roover, D., Bosgra, O.H.: Synthesis of robust multivariable iterative learning controllers with application to a wafer stage motion system. International Journal of Control 73(10), 968–979 (2000)

[37] de Roover, D., Bosgra, O.H., Steinbuch, M.: Internal-model-based design of repetitive and iterative learning controllers for linear multivariable systems. International Journal of Control **73**(10), 914–929 (2000)

[38] van der Schot, J.J., Tousain, R.L., Backx, A.C.P.M., Bosgra, O.H.: SSQP for the solution of large-scale dynamic-economic optimization problems. Computers & Chemical Engineering Supplement pp. S507–S510 (1999)

[39] Schrama, R.J.P., Bosgra, O.H.: Adaptive performance enhancement by iterative identification and control design. International Journal of Adaptive Control and Signal Processing **7**(5), 475–487 (1993)

[40] Schuurmans, J., Bosgra, O.H., Brouwer, R.: Openchannel flow model approximation for controller design. Applied Mathematical Modelling **19**(9), 525530 (1995)

[41] Schuurmans, J., van 't Hof, A., Dijkstra, S., Bosgra, O.H., Brouwer, R.: Simple water level controller for irrigation and drainage canals. Journal of Irrigation and Drainage Engineering-ASCE **125**(4), 189–195 (1999)

[42] Soetjahjo, J., Go, Y.G., Bosgra, O.H.: DIAG–a structural diagnose tool for interconnection assignment in model building and re-use. Computers & Chemical Engineering Supplement **22**, S933–S936 (1998)

[43] Steinbuch, M., de Boer, W.W., Bosgra, O.H.: Optimal-control of wind power-plants. Journal of Wind Engineering and Industrial Aerodynamics **27**(1-3), 237–246 (1988)

[44] Steinbuch, M., Bosgra, O.H.: Optimal output feedback of a wind energy conversion system. In: A. Calvaer (ed.) Power Systems Modelling and Control Applications. Selected Papers from the IFAC Symposium on Power Systems : Modelling and Control Appl., Brussels, Belgium, September 5-8, 1988, pp. 313–319. Pergamon Press, Oxford, UK (1989)

[45] Steinbuch, M., Bosgra, O.H.: Dynamic modelling of a generator/rectifier system. IEEE Trans. Power Electronics **7**(1), 212–223 (1992)

[46] Steinbuch, M., van Groos, P.J.M., Schootstra, G., Wortelboer, P.M.R., Bosgra, O.H.: Synthesis for a compact disc player. International Journal of Robust and Nonlinear Control **8**, 169–189 (1998)

[47] Steinbuch, M., Schootstra, G., Bosgra, H., O.: Robust control of a compact disc mechanism. In: W. Levine (ed.) CRC Control Handbook, section 3.2.4.4. CRC Press, Boca Raton, FL, USA (1995)

[48] Steinbuch, M., Schootstra, M., Bosgra, O.H.: Robust control of a compact disc mechanism. In: W. Levine (ed.) Control Systems Applications, pp. . 231–237. CRC Press, Boca Raton, FL, USA (2000)

[49] Steinbuch, M., Terlouw, J.C., Bosgra, O.H., Smit, S.G.: Multivariable application of μ-analysis to servo system design in the consumer electronics industry. Successful Industrial Applications of Multivariable Analysis, IEE Colloquium on pp. 8/1–8/3 (1990)

[50] Steinbuch, M., Terlouw, J.C., H., B.O.: Uncertainty modelling and structured singular-value computation applied to an electromechanical system. IEE Proceedings-D Control Theory and Applications **139**(3), 301–307 (1992)

[51] Stork, M., Tousain, R.L., Wieringa, J.A., Bosgra, O.H.: A MILP approach to the optimization of the operation procedure of a fed-batch emulsification process in a stirred vessel. Computers & Chemical Engineering **27**, 1681–1691 (2003)

[52] Tichem, M., Bosgra, O.H.: Uitdagingen voor de symbiose van mechatronica en microsystemen. Mikroniek **44**(2), 28–33 (2004)

[53] Tousain, R.L., Bosgra, O.H.: Market-oriented scheduling and economic optimization of continuous multi-grade chemical processes. Journal of Process Control **16**(3) (2006)

[54] van de Wal, M., van Baars, G., Sperling, F., Bosgra, O.H.: Multivariable $\mathcal{H}_\infty/\mu$ feedback control design for high-precision wafer stage motion. Control Engineering Practice **10**(7), 739–755 (2002)

[55] van der Weiden, A.J.J., Bosgra, O.H.: The determination of structural properties of a linear multivariable system by operations of system similarity. 1. strictly proper systems. International Journal of Control **29**(5), 835–860 (1979)

[56] van der Weiden, A.J.J., Bosgra, O.H.: The determination of structural properties of a linear multivariable system by operations of system similarity. 2. non-proper systems in generalized state-space form. International Journal of Control **32**(3), 489–537 (1980)

230

[57] van de Wijdeven, J., Bosgra, O.H.: Residual vibration suppression using Hankel iterative learning control. International Journal of Robust and Nonlinear Control 18(10), 1034–1051 (2008)

[58] Wortelboer, P.M.R., Steinbuch, M., Bosgra, O.H.: Iterative model and controller reduction using closed-loop balancing, with application to a compact disc mechanism. International Journal of Robust and Nonlinear Control 9, 123–142 (1999)

[59] Zandvliet, M.J., Bosgra, O.H., Jansen, J.D., Van den Hof, P.M.J., Kraaijevanger, J.F.B.M.: Bang-Bang control and singular arcs in reservoir flooding. Journal of Petroleum Science and Engineering 58, 186–200 (2007)

[60] Zandvliet, M.J., Van Doren, J.F.M., Bosgra, O.H., Jansen, J.D., Van den Hof, P.M.J.: Controllability, observability and identifiability in single-phase porous media flow. Computational Geosciences 12, 605–622 (2008)

[61] van Zee, G.A., Bosgra, O.H.: Gradient computation in prediction error identification of linear discrete-time systems. IEEE Transactions on Automatic Control 27(3), 738–739 (1982)

[62] van Zee, G.A., Schinkel, W.M.M., Bosgra, O.H.: Estimation of the transfer-function, time moments, and model parameters of a flow process. AICHE Journal 33(2), 341–346 (1987)

Conference Papers

[1] Aling, H., Bosgra, O.H.: Structural identifiability conditions for systems operating in closed loop. In: M.A. Kaashoek, A.C.M. Ran, J.H. van Schuppen (eds.) Realization and Modelling in System Theory. Birkhäuser Boston Inc., 1990. Proc. Int. Symp. Mathem. Theory Netw. Syst. (MTNS'89). Amsterdam (1990)

[2] Aling, H., Bosgra, O.H.: Identification of state space systems in closed loop. In: Proc. 1991 American Control Conference, pp. 1764–1769. Boston, Massachusetts (1991)

[3] Backx, T., Bosgra, O.H., Marquardt, W.: Integration of model predictive control and optimization of processes. In: Proc. ADCHEM international symposium on advanced control of chemical processes, pp. 249–260 (2000)

[4] Barton, K.L., van de Wijdeven J. J. M. Alleyne, A.G., Bosgra O. H., S.M.: Norm optimal cross-coupled Iterative Learning Control. In: Proc. 47th IEEE Conf. Decision and Control, pp. 3020–3025. Cancun, Mexico (2008)

[5] Bongers, P.M.M., Bosgra, O.H.: Low order robust h_∞ controller synthesis. In: Proc. 29th IEEE Conf. Decision and Control. Honolulu, Hawai (1990)

[6] Bongers, P.M.M., Dijkstra, S., Bosgra, O.H.: Modelling and control of flexible wind turbines. In: Proc. 2nd European Community Wind Energy Conference, pp. 516–520. Madrid, Spain (1990)

[7] Bongers, P.M.M., Steinbuch, M., Bosgra, O.H.: L_∞ calculation for generalized state space systems in continuous and discrete time. In: Proc. 1991 American Control Conference, pp. 1637–1639. Boston, Massachusetts (1991)

[8] Bosgra, O.H.: On the structure and parametrization of non minimal partial realizations. In: Proc. Int. Symposium Mathem. Theory Netw. and Systems. Beer Sheva, Israel (1983)

[9] Bosgra, O.H.: The future of mechanical engineering. In: Proc. Symposium Visions on Advanced Mechatronic Systems and Control, pp. . 7–9. Delft University of Technology, Delft, The Netherlands (2001)

[10] Dijkstra, B.G., Bosgra, O.H.: Convergence design considerations of low order Q-ILC for closed loop systems, implemented on a high precision wafer stage. In: Proc. 41st IEEE Conf. Decision and Control (CD-ROM), pp. 2494–2499 (2002)

[11] Dijkstra, B.G., Bosgra, O.H.: Extrapolation of optimal lifted system ILC solution, with application to a waferstage. In: Proc. 2002 American Control Conference, pp. 2595–2600 (2002)

[12] Dijkstra, B.G., Bosgra, O.H.: Noise suppression in buffer-state iterative learning control, applied to a high precision wafer stage. In: Proc. 2002 IEEE International Conference on Control Applications and International Symposium on Computer Aided Control Systems Design, pp. 998–1003 (2002)

[13] Dijkstra, B.G., Bosgra, O.H.: Exploiting iterative learning control for input shaping, with application to a wafer. In: Proc. 2003 American Control Conference (2003)

[14] Dijkstra, B.G., Rambaratsingh, N.J., Bosgra, O.H., Steinbuch, M., Kerssemakers, S., Scherer, C.W.: Input design for optimal discrete time pointtopoint motion of an industrial xyposition table. In: Proc. 39th IEEE Conf. Decision and Control (CD-ROM), pp. 901–906. Sydney, Australia (2000)

[15] van Donkelaar, E.T., Bosgra, O.H., Van den Hof, P.M.J.: Constrained model predictive control with on-line input parametrization. In: Proc. 38th IEEE Conf. Decision and Control, pp. 3718–3721. Phoenix, Arizona USA (1999)

[16] van Donkelaar, E.T., Bosgra, O.H., Van den Hof, P.M.J.: Model predictive control with generalized input parametrization. In: Proc. European Control Conference ECC'99 (CD-ROM), pp. 1–6 (1999)

[17] Donkers, M., van de Wijdeven, J.J.M., Bosgra, O.H.: A design approach for noncausal robust Iterative Learning Control using worst case disturbance optimisation. In: Proc. 2008 American Control Conference, pp. 4567–4572. Seattle, WA, USA (2008)

[18] Donkers, M., van de Wijdeven, J.J.M., Bosgra, O.H.: Robustness against model uncertainties of norm optimal Iterative Learning Control. In: Proc. 2008 American Control Conference, pp. 4561–4566. Seattle, WA, USA (2008)

[19] van Doren, J., Van den Hof, P.M.J., Jansen, J.D., Bosgra, O.H.: Determining identifiable parameterizations for large-scale physical models in reservoir engineering. In: Proc. 17th IFAC World Congress, pp. 11,421–11,426. Seoul, Korea (2008)

[20] Dötsch, H., Van den Hof, P.M.J., Bosgra, O.H., Steinbuch, M.: Performance enhancement on the basis of identified model uncertainty sets with application to a CD mechanism. In: Proc. 14th IFAC World Congress, Nonlinear System II, Optimal Control, pp. . 97–102. Beijing, PRC (1999)

[21] Douma, S.G., Van den Hof, P.M.J., Bosgra, O.H.: Controller tuning freedom under plant identification uncertainty: double Youla beats gap in robust stability. In: Proc. 2001 American Control Conference, pp. 3153–3158 (2001)

[22] Douma, S.G., Van den Hof, P.M.J., Bosgra, O.H.: Controller tuning freedom under plant identification uncertainty: double Youla beats gap in robust stability. In: Proc. 15th IFAC World Conference (2003)

[23] Eek, R.A., Bosgra, O.H.: Modeling and observation of CSD dynamics in a continuous crystallizer. In: Prepr. First Separations Vidions Topical Conf. on Separations Technologies: New Developments and Opportunities. AIChE Annual Meeting, pp. 653–659. Miami Beach (1992)

[24] Eek, R.A., Bosgra, O.H.: Controllability of particulate processes in relation to the sensor characteristics. In: Proc. Control of Particulate Processes IV. Kananaskis Village, Alberta, Canada (1995)

[25] Eek, R.A., Bosgra, O.H.: Multivariable control of a continuous crystallization process using experimental inputoutput models. In: Proc. DYCORD+95, 4th IFAC Symp. Dynamics and Control of Chemical Reactors, Distillation Columns and Batch Processes, p. 305310. Denmark (1995)

[26] van Eijk, J., Bosgra, O.H.: Mechatronics: cooperation in business. In: Proc. Symposium Visions on Advanced Mechatronic Systems and Control, pp. 5–6. Delft University of Technology, Delft, The Netherlands (2001)

[27] van Essen, G.M., Zandvliet, M.J., Van den Hof, P.M.J., Bosgra, O.H., Jansen, J.D.: Robust optimization of oil reservoir flooding. In: Proc. 2006 IEEE International Conference on Control Applications, pp. 699–704. Munich, Germany (2006)

[28] van Essen, G.M., Zandvliet, M.J., Van den Hof, P.M.J., Bosgra, O.H., Jansen, J.D.: Robust waterflooding optimization of multiple geological scenarios. In: P. O'Dell, Al-Khatib (eds.) Proc. Annual Technical Conference and Exhibition Focus on the Future, pp. . 1–7. San Antonio, Texas USA (2006)

[29] van Groos, P.M.J., Steinbuch, M., Bosgra, O.H.: Multivariable control of a compact disc player using μ-synthesis. In: Proc. 2nd European Control Conference, pp. 981–985. Groningen, Netherlands (1993)

[30] Groot Wassink, M.B., Bosch, N.J.M., Bosgra, O.H., Koekebakker, S.H.: Enabling higher jet frequencies for an inkjet printhead using iterative learning control. In: Proc. IEEE Conference on Control Applications, pp. 791–796 (2005)

[31] Groot Wassink, M.B., Bosgra, O.H., Koekebakker, S.H.: Minimization of cross-talk for an inkjet printhead using MIMO ILC. In: Proc. 2006 American Control Conference, pp. 964–969. Minneapolis, Minnesota USA (2006)

[32] Groot Wassink, M.B., Bosgra, O.H., Koekebakker, S.H., Slot, M.: Minimizing resedual vibrations and cross-talk for inkjet printheads using ILC designed simplified actuation pulses. In: Proc. International Conference on Digital Printing Technologies, pp. 69–74. Denver, Colorado (2006)

[33] Groot Wassink, M.B., Bosgra, O.H., Rixen, D.J., Koekebakker, S.H.: Modeling of an inkjet printhead for Iterative Learning Control using bilaterally coupled multiports. In: Proc. 44th IEEE Conf. on Decision and Control and the 2005 European Control Conference, pp. 4766–4772 (2005)

[34] Groot Wassink, M.B., Bosgra, O.H., Slot, M.: Enabling higher jetting frequencies for inktjet printheads using iterative learning control. In: Proc. 21st International Conference on Digital Printing Technologies (NIP21), pp. . 273–277 (2005)

[35] Groot Wassink, M.B., Zollner, F., Bosgra, O.H., Koekebakker, S.H.: Improving the drop-consistency of an inkjet printhead using meniscus-based iterative learning control. In: Proc. 2006 IEEE International Conference on Control Applications, pp. 2830–2835. Munich, Germany (2006)

[36] Hakvoort, R.G., Van den Hof, P.M.J., Bosgra, O.H.: Consistent parameter bounding identification using cross-covariance constraints on the noise. In: Proc. 32nd IEEE Conf. Decision and Control, pp. 2601 – 2606. San Antonio, TX, USA (1993)

[37] van Helvoirt, J., Bosgra, O.H., de Jager, B., Steinbuch, M.: Approximate realization of valve dynamics with time delay. In: Proc. 16th IFAC World Congress. Prague, Czech Republic (2005)

[38] van Helvoirt, J., Bosgra, O.H., de Jager, B., Steinbuch, M.: Approximate realization with time delay. In: Proc. 2005 IEEE Conference on Control Applications, pp. 1534–1539. Toronto, Canada (2005)

[39] van Hessem, D.H., Bosgra, O.H.: Closed-loop stochastic dynamic process optimization under input and state constraints. In: Proc. 2002 American Control Conference, pp. 2023–2028 (2002)

[40] van Hessem, D.H., Bosgra, O.H.: A conic reformulation of model predictive control including bounded and stochastic disturbances under state and input constraints. In: Proc. 41st IEEE Conf. Decision and Control (CD-ROM), pp. 4643–4648 (2002)

[41] van Hessem, D.H., Bosgra, O.H.: A full solution to the constrained stochastic closed-loop MPC problem via state and innovations feedback and its receding horizon implementation. In: Proc. 42nd IEEE Conf. Decision and Control, pp. 929–934 (2003)

[42] van Hessem, D.H., Bosgra, O.H.: Towards a separation principle in closed-loop predictive control. In: Proc. 2003 American Control Conference, pp. 4299–4304. Denver (2003)

[43] van Hessem, D.H., Bosgra, O.H.: Closed-loop stochastic model predcitive control in a receding horizon implementation on a continuous polymerization reactor example. In: Proc. 2004 American Control Conference, pp. 914–919 (2004)

[44] van Hessem, D.H., Bosgra, O.H., Scherer, C.W.: LMI-based closed-loop economic optimization of stochastic process operation under state and input constraints. In: Proc. 40th IEEE Conf. on Decision and Control, pp. 4228–4233 (2001)

[45] Heuberger, P.S.C., Bosgra, O.H.: Approximate system identification using system based orthonormal functions. In: Proc. 29th IEEE Conf. Decision and Control, pp. 1086–1092. Honolulu, Hawai (1990)

[46] Heuberger, P.S.C., Van den Hof, P.M.J., Bosgra, O.H.: A generalized orthonormal basis for linear dynamical systems. In: Proc. 32nd IEEE Conf. Decision and Control, pp. 2850 – 2855. San Antonio, TX, USA (1993)

[47] Heuberger, P.S.C., Van den Hof, P.M.J., Bosgra, O.H.: Modelling linear dynamical systems through generalized orthonormal basis functions. In: G. Goodwin, R. Evans (eds.) Proc. 12th Triennial World IFAC World Congress, vol. 5, p. 19. Pergamon Press, Oxford, UK, Sydney, Australia (1995)

[48] Van den Hof, P.M.J., Schrama, R.J.P., Bosgra, O.H.: An indirect method for transfer function estimation from closed loop data. In: Proc. 31st IEEE Conf. Decision and Control, pp. 1702–1706. Tucson, AZ, USA (1992)

[49] Van den Hof, P.M.J., Schrama, R.J.P., Bosgra, O.H.: Identification of normalized coprime plant factors for iterative model and controller enhancement. In: Proc. 32nd IEEE Conf. Decision and Control, pp. 2839 – 2844. San Antonio, TX, USA (1993)

[50] Hol, C.W.J., Scherer, C.W., van der Meché, E.G., Bosgra, O.H.: A trust-region interior point approach to fixed-order control applied to the benchmark system. In: Proc. International Workshop on Design and Optimisation of Restricted Complexity Controllers, pp. 1–6 (2003)

[51] Huesman, A.E.M., Bosgra, O.H., Van den Hof, P.M.J.: Degrees of freedom analysis of economic dynamic optimal plantwide operation. In: J. Moreno, B. Foss, M. Perrier (eds.) Proc. 10th International Symposium on Computer Applications in Biotechnology and the 8th International Symposium on Dynamics and Control of Process Systems, CAB 2007/Dycops 2007, pp. . 165–170. Cancun, Mexico (2007)

[52] Kadam, J.V., Schlegel, M., Marquardt, W., Tousain, R.L., van Hessem, D.H., van den Berg, J., Bosgra, O.H.: A two-level strategy of integrated dynamic optimization and control of industrial processes: a case study. In: J. Grievink, J. van Schijndel (eds.) European Symposium on Computer Aided Process Engineering, pp. 511–516 (2002)

[53] Kalbasenka, A.N., Huesman, A.E.M., Kramer, H.J.M., Bosgra, O.H.: Controllability analysis of industrial crystallizers. In: Proc. 12th International Workshop on Industrial Crystallization (BIWIC), pp. 157–164 (2005)

[54] Kalbasenka, A.N., Huesman, A.E.M., Kramer, H.J.M., Bosgra, O.H.: On improved experiments for estimation of Gahn's kinetic parameters. In: S. Jiang, P. Jansens, J. ter Horst (eds.) Proc. 13th International Workshop on Industrial Crystallization, pp. 122–129. Delft, The Netherlands (2006)

[55] Lambrechts, P.F., Bosgra, O.H.: The parametrization of all controllers that achieve output regulation and tracking. In: Proc. 30th IEEE Conf. Decision and Control, pp. 569–574. Brighton, UK (1991)

[56] Leskens, M., van Kessel, J.L.F., Van den Hof, P.M.J., Bosgra, O.H.: Nonlinear model predictive control with moving horizon state and disturbance estimation-with application to MSW combustion. In: Proc. 16th IFAC World Congress, pp. 1–6 (2005)

[57] Leskens, M., van Kessel, L.B.M., Bosgra, O.H.: Model predictive control as a tool for improving the process operation performance of MSW combustion plants. In: M. Carvalho, F. Lockwood, W. Fiveland, W. McLean, J.T. de Azevedo (eds.) Proc. 7th International Conference on Energy for a Clean Environment (2003)

[58] van der Meché, E., Bosgra, O.H.: A convex relaxation approach to real rational frequency domain identification. In: Proc. 41st IEEE Conf. Decision and Control (CD-ROM), pp. 252–257 (2002)

[59] van der Meché, E., Bosgra, O.H.: A convex relaxation approach to multivariable state-space frequency response approximation. In: Proc. 2003 American Control Conference, pp. 5274–5279. Denver (2003)

[60] van der Meché, E.G., Bosgra, O.H.: Nevanlinna-Pick interpolation with degree constraint: complete parameterization based on Lyapunov inequalities. In: Proc. 43rd IEEE Conf. Decision and Control, pp. 411–416 (2004)

[61] van der Meulen, S.H., Tousain, R.L., Bosgra, O.H.: Fixed structure feedforward controller tuning exploiting iterative trials, applied to a high-precision electromechanical servo system. In: Proc. 2007 American Control Conference, pp. 4033–4039. New York City, NY, USA (2007)

[62] van der Meulen, S.H., Tousain, R.L., Bosgra, O.H., Steinbuch, M.: Machine-in-the-loop control optimization. In: M. Heertjes (ed.) Proc. 8th Philips Conference on Applications of Control Technology, p. 119125. Hilvarenbeek, Netherlands (2005)

[63] Molenaar, D.P., Bosgra, O.H., Hoeijmakers, M.J.: Identification of synchronous generator transfer functions from standstill test data. In: Proc. 2002 ASME Wind Energy Symposium, pp. 331–339 (2002)

[64] Molenaar, D.P., Hoeijmakers, M.J., Bosgra, O.H.: Time-domain identification of the transfer functions of Park's DQ-axis synchronous generator model from standstill data. In: P. Helm,

A.Z. Proceedings (eds.) Proc. European Wind Energy Conference and Exhibition (Copenhagen, 2-6 July 2001),WIP, Munich, 2001, pp. 1026–1029 (2001)

[65] Nuij, P.W.J.M., Steinbuch, M., Bosgra, O.H.: Bias removal in higher order sinusoidal input describing functions. In: IEEE International Instrumentation and Measurement Technology Conference Proceedings, IMTC 2008, pp. 1653–1658. Victoria, BC, Canada (2008)

[66] Oomen, T.A.E., Bosgra, O.H.: Dealing with flexible modes in 6-DOFs robust servo control. In: .G.Z. Angelis (ed.) Proc. 9th Philips Conference on Applications of Control Technology, pp. 133–134. Hilvarenbeek, The Netherlands (2007)

[67] Oomen, T.A.E., Bosgra, O.H.: Estimating disturbances and model uncertainty in model validation for robust control. In: Proc. 47th IEEE Conf. Decision and Control, pp. 5513–5518. Cancun, Mexico (2008)

[68] Oomen, T.A.E., Bosgra, O.H.: Robust-control-relevant coprime factor identification: A numerically reliable frequency domain approach. In: Proc. 2008 American Control Conference, pp. 625–631. Seattle, WA, United States (2008)

[69] Oomen, T.A.E., van de Wal, M., Bosgra, O.H.: Exploiting H_∞ sampled-data control theory for high-precision electromechanical servo control design. In: Proc. 2006 American Control Conference, pp. 1086–1091. Minneapolis, MN, USA (2006)

[70] Oomen, T.A.E., van de Wal, M., Bosgra, O.H.: Aliasing of resonance phenomena in sampled-data feedback control design: Hazards, modeling, and a solution. In: Proc. 2007 American Control Conference, pp. 2881–2886. New York City, NY, USA (2007)

[71] Oomen, T.A.E., van de Wal, M., Bosgra, O.H., Steinbuch, M.: Optimal digital control of analog systems: A survey. In: M. Heertjes (ed.) Proc. 8th Philips Conference on Applications of Control Technology, pp. 139–144. Hilvarenbeek, Netherlands (2005)

[72] Oomen, T.A.E., van de Wijdeven, J.J.M., Bosgra, O.H.: Suppressing intersample behavior in Iterative Learning Control. In: Proc. 47th IEEE Conf. Decision and Control, pp. 2391–2397. Cancun, Mexico (2008)

[73] van Oosten, C.L., Bosgra, O.H., Dijkstra, B.G.: Reducing residual vibrations through iterative learning control with application to a wafer stage. In: Proc. 2004 American Control Conference, pp. 5150–5155 (2004)

[74] Parra, M., Bosgra, O.H., Valk, P.: Informative experiment design and non-parametric identification of 6-DOF motion platform. In: Proc. 16th IFAC World Congress, pp. 1–6 (2005)

[75] de Roover, D., Bosgra, O.H.: Dualization of the internal model principle in compensator and observer theory with application to repetitive and learning control. In: Proc. 1997 American Control Conference, vol. 6, p. 39023906. Albuquerque, New Mexico (1997)

[76] de Roover, D., Bosgra, O.H.: An internal-model-based framework for the analysis and design of repetitive and learning controllers. In: Proc. 36th IEEE Conf. Decision and Control, p. 37653770. San Diego, CA (1997)

[77] de Roover, D., Sperling, F.B., Bosgra, O.H.: Point-to-point control of a MIMO servomechanism. In: Proc. 1998 American Control Conference, p. 26482651. Philadelphia, PA, USA (1998)

[78] Schlegel, M., Backx, T., Bosgra, O.H., Brouwer, P.J.: Towards integrated dynamic real-time optimalization and control of industrial processes. In: Proc. Foundations of Computer-Aided Process Operations (2003)

[79] Schrama, R.J.P., Bongers, P.M.M., Bosgra, O.H.: Robust stability under simultaneous perturbations of linear plant and controller. In: Proc. 31st IEEE Conf. Decision and Control, pp. 2137–2139. Tucson, AZ, USA (1992)

[80] Steinbuch, M., Bosgra, O.H.: Fixed order H_2 / H_∞ control of uncertain systems. In: D. Franke, F. Kraus (eds.) Proc. IFAC Symposium on Design Methods of Control Systems. Zurich (1991)

[81] Steinbuch, M., Bosgra, O.H.: Mixed H_2 / H_∞ optimal control of a flexible mechanical servo system. In: International Conference Control '91, Edinburgh, UK, March 25-28, 1991. IEE Conference Publication number 332, pp. 184–189 (1991)

[82] Steinbuch, M., Bosgra, O.H.: Necessary conditions for static and fixed order dynamic H_2 / H_∞ optimal control. In: Proc. 1991 American Control Conference, pp. 1137–1142. Boston, Massachusetts (1991)

[83] Steinbuch, M., Bosgra, O.H.: Robust linear quadratic output feedback of a 5d electro-mechanical actuator. In: H. Barker (ed.) Prepr. 5th IFAC/IMACS Symposium on Computer Aided Design in Control Systems, pp. 437–442. Swansea, UK (1991)

[84] Steinbuch, M., Bosgra, O.H.: Robust performance in H_2 / H_∞ optimal control. In: Proc. 30th IEEE Conf. Decision and Control, pp. 549–550. Brighton, UK (1991)

[85] Steinbuch, M., Bosgra, O.H.: h_2 performance for unstructured uncertainties. In: I. Koch (ed.) Proc. IMACS Symposium on Mathematical Modelling MATHMOD, pp. 65–268. Vienna, Austria (1994)

[86] Steinbuch, M., Bosgra, O.H.: Robust performance H_2 / H_∞ optimal control. In: Proc. 33rd IEEE Conf. Decision and Control, Lake Buena Vista, FL. December 14-16, pp. 3167-3172 (1994)

[87] Steinbuch, M., van Groos, P., Schootstra, G., Bosgra, O.H.: Multivariable control of a Compact Disc Player using DSPs. In: Proc. 1994 American Control Conference, pp. 2434–2438 (1994)

[88] Steinbuch, M., J.C., T., Bosgra, O.H., Smit, S.G.: Parametric uncertainty modelling and real and complex structured singular value computation applied to an electro-mechanical system. In: Proc. International Conference Control '91, Edinburgh, UK, March 25-28, 1991. IEE Conference Publication number 332, pp. 417–422 (1991)

[89] Steinbuch, M., Schootstra, G., Bosgra, O.H.: Robust control of a Compact Disc Player. In: Proc. 31st IEEE Conf. Decision and Control, pp. 2596–2600. Tucson, AZ, USA (1992)

[90] Steinbuch, M., Schootstra, G., Goh, H., Bosgra, O.H.: Closed loop l_1 scaling for fixed-point digital control implementation. In: Proc. 1994 American Control Conference, pp. 1157–1161 (1994)

[91] Steinbuch, M., Smit, S., Schootstra, G., Bosgra, O.H.: μ-synthesis of a flexible mechanical servo system. In: Proc. 1991 American Control Conference, pp. 593–598. Boston, Massachusetts (1991)

[92] Steinbuch, M., Terlouw, J., Bosgra, O.H.: Robustness analysis for real and complex perturbations applied to an electro-mechanical system. In: Proc. 1991 American Control Conference, pp. 556–561. Boston, Massachusetts (1991)

[93] Steinbuch, M., Wortelboer, P.M.R., van Groos, P.J.M., Bosgra, O.H.: Limits of implementation - a CD player control case study. In: Proc. 1994 American Control Conference, pp., 3209–3213 (1994)

[94] Stolte, J., Vissers, J., Backx, T., Bosgra, O.H.: Modeling local/periodic temperature variation in catalytic reactions. In: Proc. IEEE International Conference on Control Applications, CCA 2008, pp. 73–78 (2008)

[95] Stork, M., Tousain, R.L., Bosgra, O.H.: A MILP approach to the optimization of the operation policy of an emulsification process. In: J. Grievink, J. van Schijndel (eds.) European Symposium on Computer Aided Process Engineering - 12, pp. 973–978 (2002)

[96] Swaanenburg, H.A.C., Schinkel, W.M.M., van Zee, G.A., Bosgra, O.H.: Practical aspects of industrial multivariable process identification. In: H.A. Barker, P.C. Young (eds.) Proc. 7th IFAC/IFORS Symp. Identification and System Parameter Estimation, pp. 201–206. York, UK. (1985)

[97] Tousain, R.L., Boissy, J.C., Norg, M.L., Steinbusch, P.J., Bosgra, O.H.: Suppressing non-periodically repeating disturbances in mechanical servo systems. In: Proc. 37th IEEE Conf. Decision and Control, pp. 2541–2542 (1998)

[98] Tousain, R.L., Bosgra, O.H.: Efficient dynamic optimization for nonlinear model predictive control - application to a high-density poly-ethylene grade change problem. In: Proc. 39th IEEE Conf. Decision and Control (CD-ROM), pp. 1–6. Sydney, Australia (2000)

[99] Tousain, R.L., Bosgra, O.H.: A successive mixed integer linear programming approach to the grade change optimization problem for continuous chemical processes. In: J. Grievink, J. van Schijndel (eds.) European Symposium on Computer Aided Process Engineering - 12, pp. 811–816 (2002)

[100] Tousain, R.L., Hol, C.W.J., Bosgra, O.H.: Measured variable selection and nonlinear state estimation for MPC, with application to a high-density poly-ethylene reactor. In: Proc. Process Control and Instrumentation Conference. Glasgow, Scotland (2000)

236

[101]   Tousain, R.L., van der Meché, E., Bosgra, O.H.: Design strategy for iterative learning control based on optimal control. In: Proc. 40th IEEE Conf. Decision and Control (CD-ROM), pp. 4463–4468 (2001)

[102]   Tousain, R.L., Prinssen, W.J., Bosgra, O.H.: Scheduling continuous multi-product processes in compliance with process dynamics and changing market conditions. In: Proc. Second Conference on Management and Control of Production and Logistics, pp. 1–6. Grenoble, France (2000)

[103]   Toussain, R.L., Prinssen, W.J.., Bosgra, O.H.: An integrated view on plant and market dynamics: scheduling poly-ethylene production. In: Proc. Process Control and Instrumentation Conference. Glasgow, Scotland (2000)

[104]   van der Wal, M., van Baars, G., Sperling, F., Bosgra, O.H.: Experimentally validated multivariable feedback controller design for a high-precision wafer stage. In: Proc. 40th IEEE Conf. Decision and Control (CD-ROM), pp. 1583–1588 (2001)

[105]   Wassink, M.B., Bosgra, O.H., Koekebakker, S.H.: Enhancing inkjet printhead performance by MIMO Iterative Learning Control. In: Proc. 2007 American Control Conference, pp. 5472–5477. New York City, NY, USA (2007)

[106]   van de Wijdeven, J.J.M., Bosgra, O.H.: Residual vibration suppression using Hankel Iterative Learning Control. In: Proc. 2006 American Control Conference, pp. 1778–1783. Minneapolis, MN, USA (2006)

[107]   van de Wijdeven, J.J.M., Bosgra, O.H.: Stabilizability, performance, and the choice of actuation and observation time windows in Iterative Learning Control. In: Proc. 45th IEEE Conf. on Decision and Control, pp. 5042–5047 (2006)

[108]   van de Wijdeven, J.J.M., Bosgra, O.H.: Hankel Iterative Learning Control for residual vibration suppression with MIMO flexible structure experiments. In: Proc. 2007 American Control Conference, pp. 4993 – 4998. New York City, NY, USA (2007)

[109]   van de Wijdeven, J.J.M., Bosgra, O.H.: Noncausal finite-time robust Iterative Learning Control. In: Proc. 46th IEEE Conf. Decision and Control, pp. 258 – 263. New Orleans, LA, USA (2007)

[110]   Winterling, M.W., Deleroi, W., Bosgra, O.H., Huisman, H., van Overloop, J., Tuinman, E.: Modelling and verification of electromechanical interaction in a traction drive. In: Proc. 6th European Conf. Power Electronics and Applications, vol. 2, p. 395400. Sevilla, Spain (1995)

[111]   de Wolf, S., Jager, J., Bosgra, O.H., Kramer, H.: Identification and control of a 970 liter evaporative DTB crystallizer. In: Proc. 11th Symposium on Industrial Crystallization, pp. . 89–94. Garmisch-Partenkirchen (1990)

[112]   Wortelboer, P.M.R., Bosgra, O.H.: Generalized frequency weighted balanced reduction. In: Proc. 31st IEEE Conf. Decision and Control, pp. 2848–2849. Tucson, AZ, USA (1992)

[113]   Wortelboer, P.M.R., Bosgra, O.H.: Frequency weighted closed-loop order reduction in the control design configuration. In: Proc. 33rd IEEE Conf. Decision and Control, pp. 2714–2719. Lake Buena Vista, FL (1994)

[114]   Wortelboer, P.M.R., Bosgra, O.H.: An iterative algorithm for frequencyweighted H_2norm optimal reduction and centering. In: Proc. 3rd Europ. Control Conf., p. 21712176. Rome, Italy (1995)

[115]   Zandvliet, M.J., Bosgra, O.H., Van den Hof, P.M.J., Jansen, J.D., Kraaijevanger, J.F.B.M.: Bang-Bang control in reservoir flooding. In: I.I. Aavatsmark (ed.) Proc. 10th European Conference on the Mathematics of Oil Recovery - ECMOR X, pp. . 1–9. Amsterdam, Netherlands (2006)

[116]   van Zee, G.A., Bosgra, O.H.: The use of realization theory in the robust identification of multivariable systems. In: Proc. 5th IFAC Symp. Identification and System Param. Estim., pp. 477–484. Darmstadt (1979)

[117]   van Zee, G.A., Bosgra, O.H.: Validation of prediction error identification results by multivariable control implementation. In: G. Bekey, G. Saridis (eds.) Proc. 6th IFAC Symposium Identification and System Parameter Estimation 1982, pp. 643–648. Washington DC (1982)

Index